FUNDAMENTAL STRUCTURAL STEEL DESIGN–ASD

FUNDAMENTAL STRUCTURAL STEEL DESIGN–ASD

Thomas Burns, P.E.

I T P

Delmar Publishers Inc.™

NOTICE TO THE READER

Cover by Cheri Plasse
Cover photo courtesy Bethlehem Steel Corporation

Delmar Staff:
Senior Executive Editor: Mark W. Huth
Assistant Editor: Nancy Belser
Project Editor: Elena M. Mauceri
Production Coordinator: Karen Smith
Art/Design Coordinator: Cheri Plasse

For information address Delmar Publishers Inc.
3 Columbia Circle Drive, Box 15–015
Albany, New York 12203–5015

Printed in the United States of America
Published simultaneously in Canada
by Nelson Canada,
a division of The Thomson Corporation

10 9 8 7 6 5 4 3 2   XX 99 98 97 96 95 94

Library of Congress Cataloging-in-Publication Data
Burns, Thomas.
    Fundamental structural steel design, ASD / Thomas Burns.
        p.   cm.
    Includes index.
    ISBN 0–8273–5705–2 (Text)   ISBN 0-8273-5706-0 (Solutions Manual)
    1. Building, Iron and steel.   2. Structural design.   3. Steel, Structural.   I. Title.
TA684.B89   1994
624.1'821—dc20   93-3275
                                                                                CIP

NEW AND REVISED TITLES FOR 1994

Managing Construction: The Contractual Viewpoint / COLLIER
ISBN: 0–8273–5700–1

Estimating for Commercial and Residential Construction / BENEDICT
ISBN: 0–8273–5498–3

Estimating with Timberline® Precision / REICH
ISBN: 0–8273–6002–9

Practical Surveying for Technicians / LANDON
ISBN: 0–8273–3941–0

Reinforced Concrete Technology / FRENCH
ISBN: 0–8273–5495–9

*To request more information on these publications,
contact your local bookstore, or call or write to:*

Delmar Publishers Inc.
3 Columbia Circle
P.O. Box 15015
Albany, NY 12212-5015

Phone: 1-800-347-7707 • 1-518-464-3500 • Fax: 1-518-464-0301

CONTENTS

PREFACE

This text has been written for students of civil engineering technology, building construction, architecture, and related technical disciplines who are enrolled in their initial structural steel design course. It was written to provide students, at both the associate- and bachelor-degree levels, with fundamental knowledge of structural steel design using the Allowable Stress Design (ASD) method. This text reflects the latest (1989) revisions of the ninth edition of the American Institute of Steel Construction's (AISC) *Manual of Steel Construction.*

The author has attempted to provide easy-to-read, yet extensive, coverage of the most essential topics that are normally covered in the length of a semester. As one can see from the Table of Contents, there is an absence of chapters covering extraneous subject matter or topics of minimal importance, as can be found in many texts. Chapters 1 through 3 introduce the student to the concepts of allowable stress design, the behavior of steel as a material, and the process of design. These chapters lay the groundwork for what the students (as potential designers) are trying to accomplish. After this introduction, the text begins (in Chapters 4 through 10) to focus on the basic elements of steel design. It is the author's utmost desire that students fully utilize this text as a tool to further develop their understanding of the concepts regarding the design of structural steel.

The text assumes that the student has a thorough background in statics, strength of materials, and construction practices, although it requires only a solid foundation in algebra and trigonometry. The derivations of most formulas are left to the discretion of more advanced texts on this subject, since they are typically not essential to the use of such design formulas. Although this text can be fully used without the aforementioned AISC reference, it is highly recommended that students have access to the AISC manual as a means to more completely realize the potential of this text.

The examples and exercises are presented in the system of units typically found in the United States. These examples and exercises strive to convey the fundamental ideas found in the design process. Therefore, a detailed, step-by-step procedure is given in every example to facilitate students' understanding of basic design concepts. Again, the author's emphasis on providing a tool of learning is paramount in the structure of the text.

Appendix A contains a comprehensive overview of the most recently adopted steel design philosophy, Load and Resistance Factor Design (LRFD). It is the author's intention to provide a conceptual understanding of this new philosophy, so that the student can contrast it with the major focus of the text, the Allowable Stress Design method. The author is of the opinion that a full transition to the LRFD method is still years away and believes that the fundamental design concepts learned through the allowable stress method will be utilized in a better understanding of the LRFD method. In essence, the allowable stress philosophy may be viewed as a gateway to achieving a fuller comprehension of LRFD.

In closing, there are many people whom I wish to thank for their help in making this book possible. Among them are Patrick Newmann, P.E. (AISC), R. Donald Murphy (SJI), and Arthur Roth (Bethlehem Steel) for the many photographs and design charts that make the text "come to life" for the many students who have not been exposed to steel construction. I would also like to thank the members of the CET faculty at Cincinnati Technical College for exposing me to the highest degree of professionalism and excellence. Also, my gratitude is extended to my editor, Mark Huth, and assistant editor, Nancy Belser, for their work in making this text a reality. Finally, I would like to thank my wife, Kim, for her encouragement and belief that this project could indeed happen.

ACKNOWLEDGMENTS

The following persons reviewed the manuscript and offered their suggestions and guidance. Their contribution is appreciated.

Robert E. Crosland
School of Building Construction
University of Florida

R.S. Davenport
Sandhills Community College

Kenneth F. Dunker, P.E.
Dept. of Civil and Construction Engineering
Iowa State University

Irving Engel
School of Architecture
Washington University

Gregory W. Mills, P.E.
Civil Engineering Technology Coordinator
Western Kentucky University

Jack Roberts, Ph.D.
Department of Technology
East Texas State University

1

INTRODUCTION TO STEEL AND STEEL DESIGN

1.1   Steel, the Material

Steel is an alloy, which is a metal made from various elements. The two main elements that comprise steel are iron and carbon. Iron is by far the main ingredient in terms of percentages, usually making up roughly 98% of all the components found in steel. Carbon typically will comprise less than 0.5% of steel but is very important because it affects the steel's strength and hardness. In mild-carbon steels, an increasing carbon content will produce stronger and harder steels. At the same time, this increase in carbon content will result in a reduction of ductility. This loss of ductility is sometimes referred to as brittleness, which is not an advantageous quality for any structural material.

Other alloying elements, such as silicon, nickel, manganese, and copper, can also be added to steel in varying amounts to enhance certain properties, such as strength, hardness, and corrosion resistance. These elements, like carbon, also comprise a very small percentage of steel's chemical composition. The American Society of Testing Materials (ASTM) specifies chemical limits of these elements for different types of steel.

The most common structural steel used in bridges and buildings is designated A36 by the ASTM. The chemical requirements for A36 steel, as specified by the ASTM, are shown in Table1–1. Many other types of steel are also commonly

Table 1-1 ASTM Chemical Requirements for A36 Steel

Product	Shapes[a]	Plates					Bars			
Thickness, in.	All	To 3/4 Incl.	Over 3/4 to 1½ Incl.	Over 1½ to 2½ Incl.	Over 2½ to 4 Incl.	Over 4	To 3/4 Incl.	Over 3/4 to 1½ Incl.	Over 1½ to 4 Incl.	Over 4
Carbon, max, %	0.26	0.25	0.25	0.26	0.27	0.29	0.26	0.27	0.28	0.29
Manganese, %	—	—	0.80–1.20	0.80–1.20	0.85–1.20	0.85–1.20	—	0.60–0.90	0.60–0.90	0.60–0.90
Phosphorus, max, %	0.04	0.04	0.04	0.04	0.04	0.04	0.04	0.04	0.04	0.04
Sulfur, max, %	0.05	0.05	0.05	0.05	0.05	0.05	0.05	0.05	0.05	0.05
Silicon, %	—	—	—	0.15–0.40	0.15–0.40	0.15–0.40	—	—	—	—
Copper, min, % when copper steel is specified	0.20	0.20	0.20	0.20	0.20	0.20	0.20	0.20	0.20	0.20

Source: Copyright ASTM. Reprinted with permission.

[a]Manganese content of 0.85–1.35%, and silicon content of 0.15–0.40%, is required for shapes over 426 lb/ft.

Table 1–2   Mechanical Properties of Structural Steels

ASTM Designation		Minimum Yield Stress, F_y	Ultimate Tensile Strength, F_u	Description
A36		36 Ksi	58–80 Ksi	Most common type of steel used in construction. All general structural purposes.
A242		42–50 Ksi	63–70 Ksi	High-strength, low-alloy steel. Better corrosion resistance, mainly used in bridges. Superseded by A709.
A588		42–50 Ksi	63–70 Ksi	Typically referred to as "weathering" steels because they are left unpainted in their final condition. High-strength, low-alloy steel. Corrosion resistance about four times greater than that of A36. Generally superseded by A709.
A709	(36) (50) (100)	36 Ksi 50 Ksi 100 Ksi	58–80 Ksi 70 Ksi 110–130 Ksi	Bridge designated steels. Quenched and tempered high-strength, low-alloy steels. Can be produced as weathering steels, having high corrosion resistance.

produced when designers consider certain properties to be necessary for a proper design. The advantageous properties of other common steels are briefly outlined in Table 1–2. A typical example of a special design condition requiring special properties may be a corrosive environment, commonly found in saltwater regions. Such a severe environment may require that a designer specify a steel such as A588, since its corrosion resistance is approximately four times better than that of A36 steel. Another common example that requires special steel properties would be the use of high-strength steel, such as A572, in long-span or multistory structures. Steels of higher strength will commonly be specified in structures where the dead load is a major portion of the total design load, because the judicious use of these steels will lead to an overall reduction of both cost and weight. In these applications, the cost savings of high-strength steel may be found not only in the reduction of actual steel weight but also in other areas such as the foundation.

Today, and in the future, designers will have at their disposal an ever-increasing number of steels to fit their special needs.

1.2   Important Design Properties of Steel

The property of steel that is most important to the understanding of design concepts is its stress–strain behavior. This behavior is best typified in the standard tensile test usually performed in most strength of materials courses. An ideal stress–strain curve for a mild-carbon structural steel is shown in Figure 1–1.

Three important regions of this curve are vital for the designer to understand. The first region is the straight-line portion of the graph between the origin and point A. This region is referred to as the **elastic range**. In the elastic range, this linear relationship implies that some increase in stress (psi) will lead to some specific, corresponding increase in strain (in./in.). The ratio of this linear relationship determines a material's **modulus of elasticity, E,** which is typically taken as 29×10^6 psi for steel. This straight-line relationship exists until the proportional limit. Soon after the proportional limit is surpassed, the material will reach its elastic limit and then its yield point. This yield point corresponds to a stress level referred to as the material's **yield stress, F_y.** The yield stress of a material is a stress level that is very important in structural design because it serves as a limiting value of a member's usefulness.

After a material reaches its yield stress, the material strains uncontrollably with practically no corresponding increase in stress. This region is typically referred to as the **plastic region,** the second important region on the curve. Once a material is stressed past its elastic limit, it can never return to its original length. This behavior is referred to as plastic deformation, and it can be disastrous for a structural member. (Can you imagine what would occur every time a heavy truck crossed a bridge if the floor beams were allowed to be plastically deformed?)

In our discussion so far, the idea of steel entering this plastic region might be

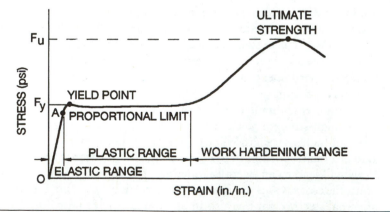

Figure 1–1   Typical Stress–Strain Diagram for Mild-Carbon Steels.

viewed as unwanted, but actually, once the steel has entered this region, plastic deformation is an important asset in any structural material. The length of the elastic and plastic region can be viewed, in essence, as a measure of the material's ductility. If a material is ductile, it will strain or deflect a large amount before actually breaking. In terms of a structural behavior, it is highly advantageous for a material to exhibit distress before it actually collapses, because of the potential warning it provides. Many structural problems can be remedied due to quick action after a member distress is first noticed. Such distress may include sagging or leaking roof systems, cracked slabs, or even displaced door and window frames.

After the ideal steel passes through its plastic region, it enters the third important region—the **work-hardening region.** In the work-hardening region, steel undergoes an ability to increase its stress-carrying capability, due to the fact that as the steel deforms, the internal structure of the metal is being dislocated.[1] These dislocations increase as the plastic deformation increases and in turn make it harder for future dislocations to occur. This concept is easily seen by trying to break a coat hanger by repeatedly bending it back and forth. Doesn't the hanger get "tougher" to bend the last two or three times before it breaks? The last important point in the stress–strain diagram is the highest stress point on the curve, the **ultimate tensile stress, F_u.** This material property is an important one, indicating the maximum stress level a member can withstand before breaking. This stress level is also used in portions of steel design.

1.3   Manufacture and Fabrication of Rolled-Steel Shapes

The term *rolled-steel* refers to the process whereby structural steel sections are manufactured today. A rolled-steel section actually starts out as a large block of red-hot steel called an ingot. This block is passed through a successive series of rollers whereby it is gradually formed into its final shape. The properties of the steel section can be further enhanced by additional processes, such as quenching and tempering, which may take place after the initial rolling. Quenching is a rapid cooling of the steel section, and tempering is the reheating of the steel to roughly 1150°F. These processes change the microstructure of steel, leading to increases in both strength and hardness.

The first steel I-beams in the United States were rolled in the late 1800s.[2] Today, a wide variety of beams and other sections are commonly rolled (Figure 1–2). These other sections include angles, channels, tees, zees, tubes, and plates. Information regarding typical rolled shapes can be obtained from the American Institute of Steel Construction (AISC), *Manual of Steel Construction.*[3] This manual gives detailed information about section shapes, design examples, design aids, and specifications. References will be made throughout this text to the ninth edition of this handbook, published in 1989, and it will be advantageous for the student to have access to this manual.

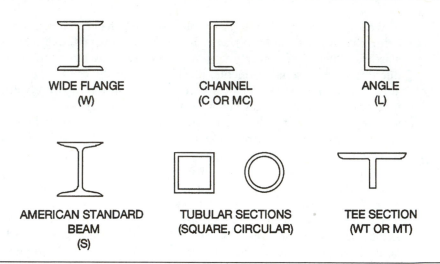

Figure 1–2   Common Rolled-Steel Shapes Used in Construction.

Since this book deals with the basics of steel design, much of the discussion will center on the wide-flange shape (W-section), which is the most common structural shape used for beams and columns. (See Figure 1–3.)

The wide-flange section is a shape with the outer edge of its flange approximately parallel to the inner edge. This facilitates the process of connecting members together with high-strength bolts. The abbreviation used when calling out a certain W-section on plans or drawings would be as follows:

$$W\ 12 \times 50$$

which means that the rolled shape is a wide-flange section, approximately 12 inches in depth, and weighs 50 pounds per foot of length. The depth and weight are particularly important pieces of information to designers because they are constantly concerned with clearance height, dead weight, and cost per pound of the structure. As students are required to know more about the cross-sectional properties of specific rolled shapes, they will be referred to the tables in the AISC manual or an applicable rendition thereof.

Manufactured steel shapes change over time, depending on demand. If there is little interest in a certain size, it may be discontinued. On the other hand, if the steel industry recognizes a demand, new standard shapes may be introduced. An example can be found by referring back to the eighth edition of the AISC's *Manual of Steel Construction*, in which the largest W-section was a W 36 × 300 at the time of publication in 1978. Today, because of demand for larger sections, there are a number of W-sections larger than a W 36 × 300, with the heaviest being a W 36 × 848.

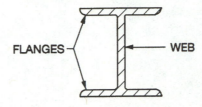

FLANGES   WEB

Figure 1–3   The Wide-Flange Section.

Fabrication of rolled steel sections and plates involves the cutting, welding, and drilling of these shapes to facilitate their erection in a steel structure. Fabrication of steel occurs in what is commonly referred to as a "fabricating shop." Such shops are involved in a variety of activities, all of which aim to prepare the raw steel shapes into their final form.

The activities of a fabricating shop include the welding together of steel plates to form "plate girders," which are used in many long-span applications, the drilling of bolt holes in their exact locations for connection purposes, and the cutting of shapes to achieve their proper length and angle of fit.

1.4   Types of Steel Structures

Because of its superior strength, quality control, and ease of fabrication, steel is an ideal material for a variety of structure types. Practically every type of structure imaginable has been (or can be) built with steel. The following paragraphs (and Figure 1–4) will try to briefly outline some of the more common structures made from steel.

Wall-bearing construction typically utilizes steel members that frame into, or sit on top of, concrete or masonry walls. The steel members are typically rolled-beam sections or, perhaps, open-web steel joists. These steel members will form the framing of a roof or floor system. A good everyday example of this type of structure is the steel beam that is pocketed into the concrete foundation wall in the basement of most houses.

Skeleton-frame construction, first used in the erection of the Home Insurance Building in Chicago in 1885, is a common steel building system. This type of construction consists of multiple numbers of floors and bays and is normally associated with multistory or office-tower construction. Steel beam members form the support systems for the individual floors, while the steel columns support these beams.

Long-span construction with steel is normally handled by construction systems employing plate girders, trusses, or arches. Plate girders are most commonly seen as bridge beams on many bridge overpasses. These beams consist of three or more

Figure 1–4(a) Skeleton Frame Construction, Georgia Railroad Bank and Trust Building, Atlanta, Georgia. (Courtesy Bethlehem Steel Corporation.)

Figure 1–4(b) Plate Girder Construction, Madison Square Garden, New York. (Courtesy Bethlehem Steel Corporation.)

Figure 1–4(c) Steel Truss Construction, Hawk Falls Bridge, Pennsylvania. (Courtesy Bethlehem Steel Corporation.)

Figure 1–4(d) Steel Arch Construction, Lewiston–Queenston Bridge, New York. (Courtesy Bethlehem Steel Corporation.)

rolled steel plates welded together to form the I-section. The depth of a plate girder will increase as the span length increases, to give the girder more moment resisting capacity. Truss-type structures are very common in bridges and long-span roof framing. A truss is made from individual steel members that, ideally, are loaded in only tension or compression. Similar to a plate girder, a truss will generally become deeper as a span length increases, to give the truss more load-carrying capability. Steel arch construction is definitely more elaborate in design and fabrication than any of the aforementioned types of steel structures and typically is relegated to bridge structures or structures dependent on aesthetics. Ideally, an arch is in full compression along its entire length, although other stresses such as bending no doubt will exist due to other factors in a real arch.

1.5   Failure in Steel Structures

When most people consider failure of a steel structure, they envision a catastrophic collapse such as the Tacoma–Narrows Bridge failure outside of Tacoma, Washington, in 1940. Practically all engineering students have seen the five-minute movie[4] that captured the failure of this immense structure, where the gently swaying superstructure rapidly turned into a violently twisted mass of steel and concrete, resulting in catastrophic collapse.

Although the early theories behind this bridge's collapse dealt with wind turbulence and resonant motion types,[5] recent arguments reject this cause, and today's debate now points to self-excitation or nonlinear behavior.[6] In either case, the bottom line is that some unknown or unseen design factor led to this failure. Collapse is not the only type of design failure that occurs; in fact, it might be the least likely in terms of frequency. Far more common is the design failure that affects a project's schedule or the serviceability of a completed structure. Items such as excessive deflection, temperature effects, or ease of erection can lead to problems that will result not in collapse but in a multitude of headaches for the designer. This leads the designer to face three key points:

1. **Serviceability** of a steel structure is most dependent upon its yield point, F_y. The stress on a member usually must not surpass this point because that will lead to permanent deformation, and the member will be considered to have failed.

2. The design of a structure must consider **all possible loading cases and other facets of the project.** Stresses imposed by wind, temperature, and seismic activity must be incorporated into the design, and the structure must mesh with architectural, mechanical, and environmental aspects of the project. To neglect these concerns will lead to delays and cost overruns, which constitute a design failure.

3. The design must be **practical.** If the structure can be built on paper only, it is of little use and has again failed due to the poor design.

Adherence to these principles will help the inexperienced designer become a success. The experience and insight of other professionals and construction personnel should always be considered and evaluated. The hardest of problems can sometimes be solved by the simplest of means.

1.6   The Allowable Stress Design Philosophy

Today, there are three generally recognized philosophies that can be used in the design of any particular material. They are listed here, followed by other commonly used names to describe the same philosophy.

1. Allowable Stress Design (elastic design, working stress design)
2. Plastic design (ultimate stress design, limit state design)
3. Load and Resistance Factor Design (LRFD)

(This book focuses on the Allowable Stress Design method and contains an overview on the Load and Resistance Factor Design method in Appendix A. The plastic design method, although legitimate, is rarely used because of its difficulty in application.)

The most common method for designing steel structures at the present time is **Allowable Stress Design (ASD)** because of its simplicity and proven track record in providing the basis for safe and reliable designs. The ASD philosophy is based on keeping the stresses in a member below some fraction of a specified stress in the steel. Referring back to the stress–strain curve for steel in Figure 1–1, it was shown that the linear region of the graph was termed the elastic region. In this region, if the load is taken off of the specimen (stress equals zero), the strain will correspondingly go back to zero (no deformation). This essentially means that the steel behaves somewhat like a rubber band; as tension is applied it stretches, and when the load is released it returns to the original length. This is the standard behavior of steel as long as you do not approach its yield point (F_y). Once the stress level goes past this point, even upon reducing the stress to zero, the member will be deformed.

The allowable stress philosophy in steel design will reduce the maximum stress levels allowed on a member to a fraction of some specified stress, such as the yield stress, F_y. These maximum stress levels, called **allowable stresses,** are given in the specifications found in the AISC manual. Such allowable levels are based on a number of factors, including type of member, type of stress, and risk of the member failure. The student should realize that the allowable levels we use today have actually evolved through years of experience and research. General allowable stresses found throughout the AISC specifications might be shown as $0.60\,F_y$, $0.66\,F_y$, or $0.75\,F_y$, depending on the behavior under consideration. It must be noted that an allowable stress given as $0.60\,F_y$ has a "factor of safety" that is the inverse of the fraction, or $1/.60 = 1.66$. This philosophy of allowable stress design actually provides a "buffer zone," or factor of safety, that extends from the fractional level

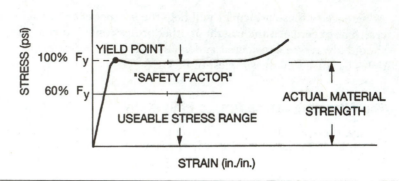

Figure 1–5   The Allowable Stress Design Philosophy (using 60% F_y as an arbitrary level.)

to the maximum the steel can take (i.e., from 0.60 F_y to 1.0 F_y; see Figure 1–5). This buffer zone is provided to guard against overload conditions, field errors, and random mistakes. Although the steel can take a much higher stress than what the specification allows, designers are limited to the allowable stress levels to provide adequate safety. This philosophy is usually conservative because a number of "bad conditions" (such as a severe overload, field mistakes, and/or material defects) would have to be present to dissipate the entire safety factor, or buffer zone.

The allowable stress method has been the primary steel design philosophy used since the introduction of steel as a building material. Although new methods of design are currently accepted as legitimate and should be explored by the student interested in steel design, ASD will no doubt survive well into our lifetimes because of its inherent safety, simplicity, and proven history.

EXERCISES

1. Investigate the uses of an American Standard beam (S-section), and list the difference between this shape and a wide-flange shape. Why is the S-section no longer used very much in modern construction?

2. List the meanings of these different designations.

    _____   C 15 × 50
    _____   L 7 × 4 × 1/2
    _____   HP 14 × 117
    _____   WT 12 × 47

3. Plot the following load-versus-deformation information for a 1/2-in.-diameter tensile specimen. The original gauge length was 2 in. Remember that the load-

versus-deformation information can be converted into stress and strain data. After plotting this information, calculate the modulus of elasticity and the approximate yield stress. What would the allowable stress be on this material if it were limited to $0.66\,F_y$?

Load (#)	Deformation (in.)
1000	.0004
1500	.0008
2000	.0012
3000	.0020
4000	.0028
5000	.0036
5500	.0047
6000	.0060
6500	.0076

REFERENCES

1. Lawrence Van Vlack, *Elements of Material Science and Engineering* (Reading, Mass.: Addison-Wesley, 1980), p. 204.
2. W. McGuire, *Steel Structures* (Englewood Cliffs, N.J.: Prentice-Hall, 1968), p. 19.
3. AISC, *Manual of Steel Construction, Allowable Stress Design,* 9th ed. (Chicago: American Institute of Steel Construction, 1989).
4. "Tacoma Narrows Bridge Collapse," Franklin Miller, Ohio State University, Department of Photography, Motion Picture division, 1963.
5. F.B. Farquharson, "Aerodynamic Stability of Suspension Bridges," University of Washington Engineering Experiment Station, Bulletin No. 116, 1941.
6. David Berreby, "The Great Bridge Controversy," *Discover,* February 1992, pp. 26–33.

2

SPECIFICATIONS, BUILDING CODES, AND TYPES OF LOADS

2.1   Specifications

Many organizations exist, both in the United States and around the world, to promote the use of a particular material in the construction market. These organizations strive to improve and advance the use of their material through extensive research into its engineering properties and behavior. The usual end result of this research is the publication of a specification for the convenience of designers and engineers.

These specifications typically contain suggested design methods that, hopefully, will ensure a safe and economical design. They also promote the information that the organization believes to be the best engineering practice at the time of publication. The evolvement of these practices will take place continually over a number of years. The prudent designer should allow these specifications to guide his or her judgment and make a serious effort to understand the principles behind the equations found in a particular specification. The judgment of a good designer is still the most important factor in design because no specification will cover every possible situation. The responsibility of designing a safe structure rests solely with the design engineer and supporting staff. A specification concerning the use of a particular material is not a magic shield for the designer to hide behind if problems occur; it simply represents the best design information that the promoting organization has to offer.

For structural steel design of buildings in the United States, the principal specification is found in the American Institute of Steel Construction's (AISC) *Manual of Steel Construction.* This manual contains not only the AISC design specifications, but also many other useful design tools such as tables and charts. Other specifications commonly used in the United States are listed. These organizations represent the best body of knowledge for design engineers to utilize in their respective areas.

> American Concrete Institute (ACI), *Building Code Requirements for Reinforced Concrete*
> American Association of State and Highway Transportation Officials (AASHTO), *Standard Specification for Highway Bridges*
> American Railway Engineering Association (AREA), *Manual for Railway Engineering*
> American Institute of Timber Construction (AITC), *Timber Construction Manual*
> American Welding Society (AWS), *Structural Welding Code*

2.2   Building Codes

Building codes, in contrast to specifications, are wide-ranging documents covering many areas of a project, such as design loadings, occupancy limits, plumbing, electrical requirements, and fire protection. Building codes are adopted by states, cities, or other governmental bodies as a legal means for protecting the public's health and welfare. The building code for a particular state, city , or county will constitute the set of rules to which all construction must adhere. Thus designers must be knowledgeable of applicable building codes as they begin a project.

Many times, a particular locality will have more than one code to which construction must conform. It is not unusual to have a project located within city limits that must conform not only to the city's building code, but also to the county and state building codes as well.

Although there is no single national building code, there are organizations whose purpose is to write "model" building codes. These model codes are sometimes integrated, fully or partially, by many state or local building codes. The most common model codes are listed next.

> Building Officials and Codes Administrators International Incorporated (BOCA), *The BOCA National Building Code*[1]
> International Conference of Building Officials, *Uniform Building Code*[2]
> Southern Building Code Congress International, *Standard Building Code*[3]

2.3   The Purpose of Design Loads

Almost all building codes provide chapters dedicated to the proper assignment and application of minimum design loads and load combination on a structure. The purpose of these chapters is to ensure that under normal conditions the structure will be both safe and serviceable. The magnitude of design loads might seem to be rather large and conservative, but the building must be designed to resist the maximum load as forecast over a certain period of time. It would be foolish to design a structure using loads that are considered to be typical everyday loads. Aren't we glad that when the wind gusts to 45 mph during a (rather typical) spring storm that structures were not designed for an everyday wind gust of 15 mph? The designer hopes the structure remains serviceable and functioning even in the event that loads reach (or exceed) these maximum values. An event such as an overload condition is a case when the size of the "buffer zone," or factor of safety (which was discussed in Chapter 1), becomes extremely important.

Many maximum design live loads used in building codes are based on a return period of 50 to 100 years, suggesting that these maximum loads would be expected to occur once every 50 or 100 years. An excellent guide to design loads and their application is the American Society of Civil Engineers (ASCE) *Minimum Design Loads for Buildings and Other Structures.*[4]

2.4   Types of Loads

Building structures are designed to resist many types of loads, such as dead loads, live loads, wind loads, snow loads, and earthquake loads. The complete design must consider all effects of these loads, both individually and in combination with one another. A short explanation of these loads and their influence in design is explored in the following paragraphs.

Dead Loads

Dead loads are all permanent loads imposed on a structure. Such loads will include the structure's self-weight, piping, conduit, permanent equipment, and permanent furniture.

These loads can typically be estimated with a high degree of accuracy in the preliminary design and should not radically change over a structure's life, unless a renovation of the structure is undertaken. In steel design, the calculation of self-weight is made even easier by the standard designations of rolled shapes. If it is known that a building frame consists of 300 lineal feet of W 12 × 87's, the calculation of the approximate weight of that steel frame is simply (300 ft × 87 lb/ft) equal to 26,100 lb. The weights and densities of various construction materials can be found in the appendix of most building codes. An abbreviated list is shown in Table 2–1.

Table 2–1 Typical Recommended Dead Loads for Construction

Densities

Steel	490 #/ft^3
Reinforced concrete	150 #/ft^3
Dry soil	90–120 #/ft^3
Brick masonry	100–130 #/ft^3
Stone Masonry	140–175 #/ft^3
Water	62.4 #/ft^3
Seasoned wood (pine)	37 #/ft^3
Seasoned wood (oak)	47 #/ft^3
Ceramic tile	150 #/ft^3
Plywood	36 #/ft^3

Weights

Asphalt shingles	2 #/ft^2
Four-ply felt and gravel composition roof	5.5 #/ft^2
20-gauge metal deck	2.5 #/ft^2
Plywood (per 1/8″ thickness)	.4 #/ft^2
7/8″ hardwood flooring	4 #/ft^2
2 × 8 floor @ 16″ spacing, unplastered	5 #/ft^2
2 × 10 floor @ 16″ spacing	6 #/ft^2
Drywall	4 #/ft^2
2 × 4 wall, unplastered	4 #/ft^2
Ceramic tile on 1″ mortar bed	23 #/ft^2

EXAMPLE 2.1

Calculate the dead load per foot on the rolled steel section as shown in the sketch. The beam in question has a 6-ft-wide distributive floor area loading it.

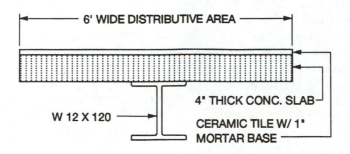

The load on the beam as shown comes from three sources: the self-weight of the beam, the 4-in.-thick concrete slab, and ceramic tile on the 1-in. mortar base. The calculation per foot can be performed as follows:

Steel beam: 120 lb per lineal foot = 120 lb/ft

Concrete slab: 6 ft wide × (4 in./12 in. per ft) × 150 lb/ft^3 = 300 lb/ft

Ceramic flooring: 6 ft wide × 23 lb/ft^2 = 138 lb/ft

Total Dead Load = 558 lb/ft

Live Loads

Live loads are simply instantaneous or movable loads produced in a structure by the occupancy of people or mobile equipment, such as furniture. Such loads can vary in magnitude, location, and duration. Actual values of live loads are hard to predict, so most building codes will assign a minimum live load for floor or roof areas (in pounds per square foot, psf) based on intended use of a building. A typical building code might assign live load values as shown in Table 2–2.

Most of these live loads have been developed as a result of practical experience over the last century, because they have proven to yield safe and serviceable designs when correctly applied. The designer should consult the local building code for a tabular listing such as that in Table 2–2. Remember that the building code will suggest these loads as the minimum, leaving the use of higher values to the discretion of the designer if he or she deems the increase necessary.

EXAMPLE 2.2

Using the same beam as in Example 2.1, calculate the total live and dead load per foot if the beam is part of the floor in the projection room of a movie theater.

The dead load will remain the same as in Example 2.1—558 lb/ft. The live load can be found in Table 2–2 as 100 psf. Therefore the live load per foot is calculated as follows:

100 lb/ft^2 × 6 ft = 600 lb/ft

Therefore the total dead and live load equals 1158 lb/ft (600 lb/ft + 558 lb/ft).

Wind Loads

Wind loads are important for all types of structures, but they are extremely important as structures become taller with respect to their lateral width. Tall and slender

Table 2–2   Minimum Suggested Live Loads

Occupancy or Use	Live Load (lb/ft²)	Occupancy or Use	Live Load (lb/ft²)
Air-conditioning (machine space)	200*	Kitchens, other than domestic	150*
Amusement park structure	100*	Laboratories, scientific	100
Attic, nonresidential		Laundries	150*
Nonstorage	25	Libraries, corridors	80*
Storage	80*	Manufacturing, ice	300
Bakery	150	Morgue	125
Balcony		Office buildings	
Exterior	100	Business machine equipment	100*
Interior (fixed seats)	60	Files (see file room)	
Interior (movable seats)	100	Printing plants	
Boathouse, floors	100*	Composing rooms	100
Boiler room, framed	300*	Linotype rooms	100
Broadcasting studio	100	Paper storage	**
Catwalks	25	Press rooms	150*
Ceiling, accessible furred	10#	Public rooms	100
Cold storage		Railroad tracks	++
No overhead system	250‡	Ramps	
Overhead system		Driveway (see garages)	
Floor	150	Pedestrian	
Roof	250	Seaplane (see hangars)	
Computer equipment	150*	Rest rooms	60
Courtrooms	50–100	Rinks	
Dormitories		Ice skating	250
Nonpartitioned	80	Roller skating	100
Partitioned	40	Storage, hay or grain	300*
Elevator machine room	150*	Telephone exchange	150*
Fan room	150*	Theaters:	
File room		Dressing rooms	40
Duplicating equipment	150*	Grid-iron floor or fly gallery	
Card	125*	Grating	60
Letter	80*	Well beams, 250 lb/ft per pair	
Foundries	600*	Header beams, 1000 lb/ft	
Fuel rooms, framed	400	Pin rail, 250 lb/ft	
Garages—trucks	§	Projection room	100
Greenhouses	150	Toilet rooms	60
Hangars	150§	Transformer rooms	200*
Incinerator charging floor	100	Vaults, in offices	250*

Source: Reprinted from ASCE, "Minimum Design Loads for Buildings and Other Structures," ASCE 7–88, with permission from the American Society of Civil Engineers.

*Use weight of actual equipment or stored material when greater.

structures, such as skyscrapers, chemical tanks, and cooling towers for power plants, are very likely to have their design influenced by wind loadings. There is evidence that low buildings in the presence of taller, adjacent buildings might have wind pressure amplified by a factor of 2.0 in some cases.[5]

The wind pressure on a structure is assumed to be of uniform intensity based on the velocity pressure (in psf) obtained from the basic wind speed, V, which is taken at a distance of 33 ft above the ground. Such velocities are taken from a chart such as that shown in Figure 2–1. This expression for wind pressure is:

$$q_z = .00256 \, K_z(IV)^2$$

where

> q_z = pressure at height z above ground, psf
> V = basic wind speed (100 year wind) taken from Figure 2–1 or similar chart
> I = structure importance factor
> K_z = velocity pressure coefficient

This basic pressure is modified by coefficients for wind gust and shape factor. The pressures, which are assumed to act uniformly, generally press against the windward wall and cause suction on the leeward wall. Pressures on the roof can be either upward or downward, depending on the roof shape, structure shape, and direction of loading. Generally, it is assumed that on roofs of steeper slopes, the wind pressure will again push on the windward side and cause suction on the leeward side.

Snow Loads

Snow loads are handled much in the same way as wind loads; that is, normally, the basic value (termed ground snow load) is taken from isolines on the map shown in Figure 2–2. This ground snow load value can then be modified by coefficients that account for exposure, structure importance, and whether the building is heated. For example, if a building is located in open terrain, the basic ground snow value

---

‡Plus 150 lb/ft² for trucks.

§Use American Association of State Highway and Transportation Officials lane loads. Also subject to not less than 100% maximum axle load.

**Paper storage 50 lb/ft of clear story height.

#As required by railroad company.

*Accessible ceilings normally are not designed to support persons. The value in this table is intended to account for occasional light storage or suspension of items. If it may be necessary to support the weight of maintenance personnel, this shall be provided for.

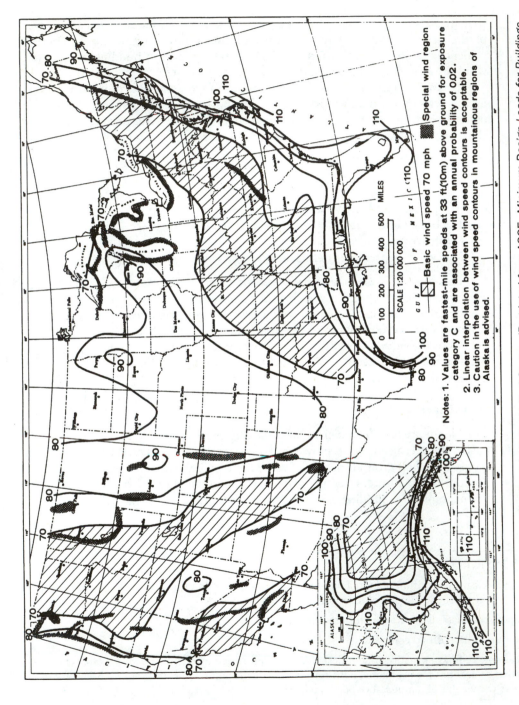

Figure 2–1 Basic Design Wind Speeds in the United States. (Reprinted from ASCE, *Minimum Design Loads for Buildings and Other Structures*, ASCE 7–88, with permission from the American Society of Civil Engineers.)

Notes: 1. Values are fastest-mile speeds at 33 ft(10m) above ground for exposure category C and are associated with an annual probability of 002.
2. Linear interpolation between wind speed contours is acceptable.
3. Caution in the use of wind speed contours in mountainous regions of Alaska is advised.

☐ Basic wind speed 70 mph   ▨ Special wind region

SCALE 1:20 000 000

0   100   200   300   400   500   MILES

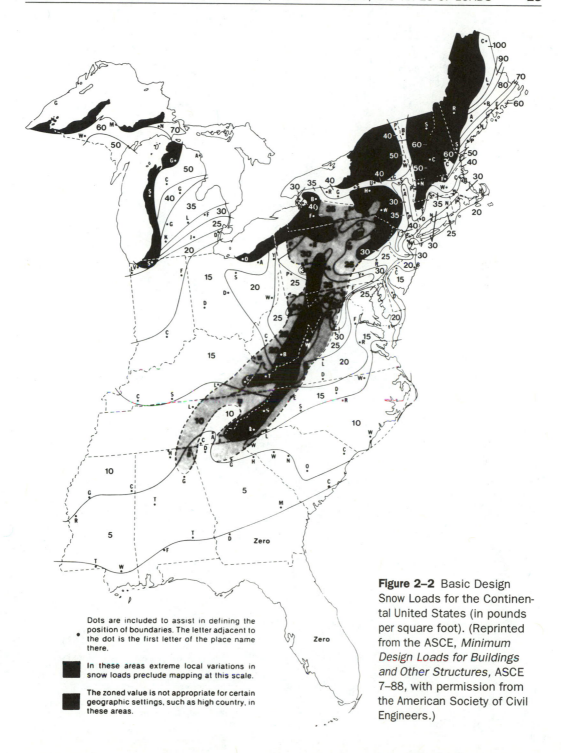

Figure 2–2 Basic Design Snow Loads for the Continental United States (in pounds per square foot). (Reprinted from the ASCE, *Minimum Design Loads for Buildings and Other Structures,* ASCE 7–88, with permission from the American Society of Civil Engineers.)

Dots are included to assist in defining the position of boundaries. The letter adjacent to the dot is the first letter of the place name there.

In these areas extreme local variations in snow loads preclude mapping at this scale.

The zoned value is not appropriate for certain geographic settings, such as high country, in these areas.

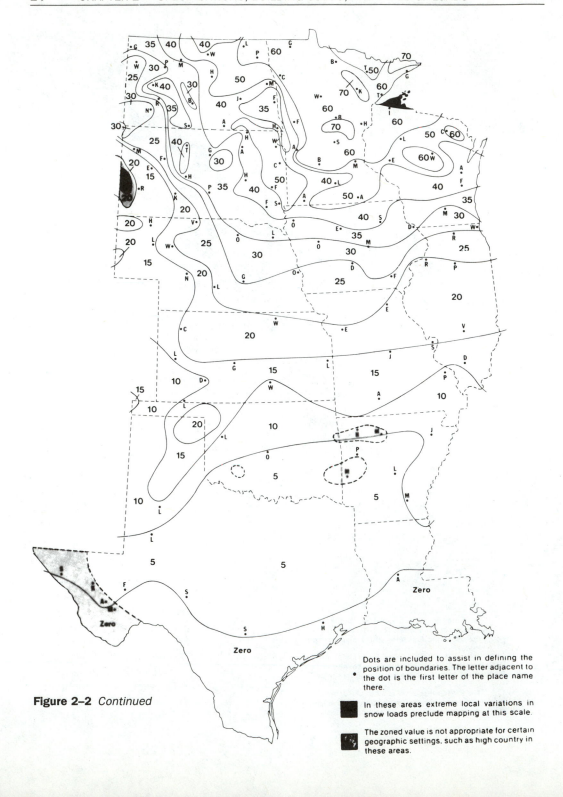

Figure 2–2 *Continued*

Dots are included to assist in defining the position of boundaries. The letter adjacent to the dot is the first letter of the place name there.

In these areas extreme local variations in snow loads preclude mapping at this scale.

The zoned value is not appropriate for certain geographic settings, such as high country in these areas.

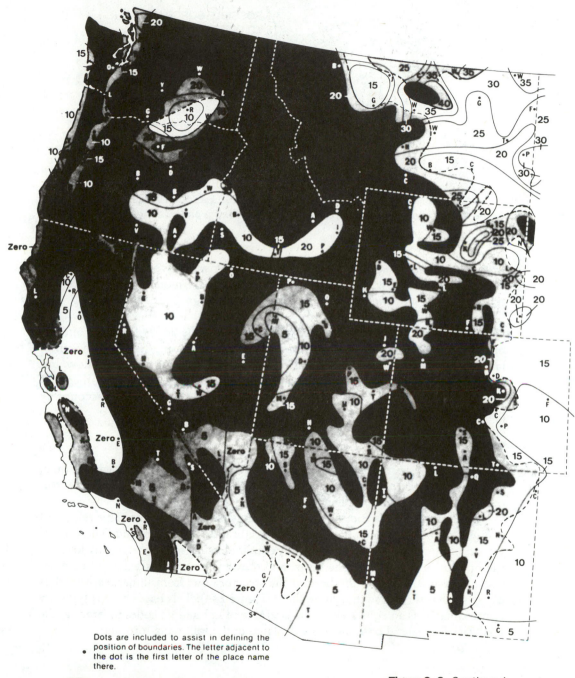

Dots are included to assist in defining the position of boundaries. The letter adjacent to the dot is the first letter of the place name there.

In these areas extreme local variations in snow loads preclude mapping at this scale.

The zoned value is not appropriate for certain geographic settings, such as high country, in these areas.

Figure 2–2 *Continued*

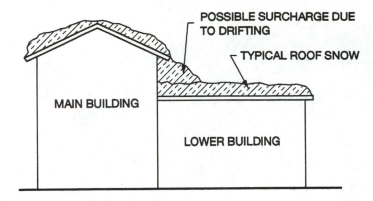

Figure 2–3   Additional Load from Potential Snow Drifts on Lower Adjacent Roofs.

would be reduced because of the probability that winds would prevent a large accumulation of snow. For more information on these modifiers, the student is referred to the aforementioned ASCE *Minimum Design Loads for Buildings and Other Structures.*

Other factors that must be taken into account are the slope of the roof and the profile of adjacent roofs. Both of these items can affect the depth of snow accumulation due to the possibility of snow drifts. Steeper roofs accumulate less snow, but lower adjacent roofs have severe snow drift potential. Parapet wall and other roof projections also can lead to severe drifting. Most codes take this into account by placing a snow "surcharge" in these particular areas (Figure 2–3).

Earthquake Loads

Although severe earthquakes have been recorded throughout the history of man, the building code requirements regarding the structural design to resist earthquakes are a rather recent phenomena.[6]

An earthquake is a sudden release of energy along a fault line. This release of energy sets off both horizontal and vertical vibrations of the ground. The vertical vibrations, or shock waves, are typically counteracted through the gravity loads of the building's self-weight. The horizontal movements are of greater concern because they tend to move the base of the structure out from underneath it, thus causing what is referred to as base shear (Figure 2–4). This base shear is typically approximated as an equivalent static lateral load in building codes to simplify the analysis. This lateral base shear is calculated in the following formula:

$$V = \frac{ZIC}{R_w} W$$

where

V = total base shear to be distributed as lateral loads over floor levels as per code.

Z = factor reflecting the possible peak ground acceleration of an earthquake occurring in a particular region, a higher number representing a greater acceleration. For example, much of California is located in Zone 4 (see Figure 2–5), which is assigned the highest acceleration coefficient in any building code.

I = factor representing the importance of the structure. Hospitals, fire stations, and other critical emergency structures are given a higher factor.

C = coefficient relating to the structure's natural frequency and soil conditions. Buildings with smaller frequencies tend not to match the frequencies of a normal earthquake, and this property is desirable because resonant behavior is avoided.

R_w = response modification factor reflecting the reduction of a structure's response due to damping and inelasticity.

W = the structure's dead weight under seismic conditions.

For a look at the coefficients assigned to each of these earthquake factors, the student is referred to his or her state building code.

The application of a static load gives only a rough approximation of the forces applied to a structure by an earthquake. Other methods, called dynamic methods, have been developed to approximate more closely the true force of an earthquake.

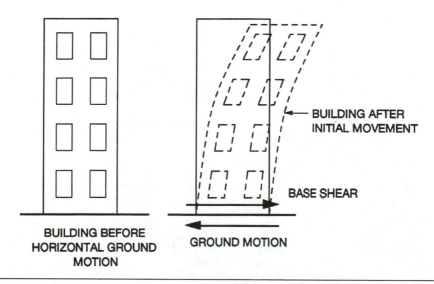

Figure 2–4   Base Shear on Buildings Under Earthquake Forces.

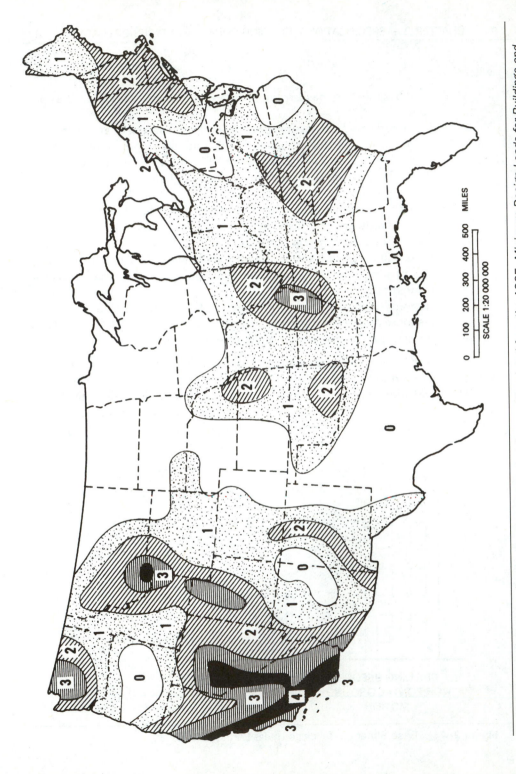

Figure 2–5 Seismic Zones for the Continental United States. (Reprinted from the ASCE, *Minimum Design Loads for Buildings and Other Structures*, ASCE 7–88, with permission from the American Society of Civil Engineers.)

SCALE 1:20 000 000

0   100   200   300   400   500

MILES

Figure 2–6   Erection of the Quinnipiac River Bridge in Connecticut Demonstrates the Importance of Construction Loads in Design. (Courtesy Bethlehem Steel Corporation.)

A common dynamic method of analysis is termed the *spectrum technique,* which focuses on the earthquake's energy applied to a building and how that building will absorb this energy. While dynamic methods are more exact with respect to the true response of the structure, they are also more rigorous to perform.

Miscellaneous Loads

Other loads that a structure has to tolerate can sometimes be overlooked by the inexperienced designer. For example, construction loads will always be present during the erection of a structure, and often these loads can be far greater than any live load placed on the structure in everyday operation. The effect of such loads can be exaggerated due to the structure's "incompleteness" during the construction phase. A good example can be seen in Figure 2–6, which shows the cantilevered portions of a plate girder bridge before erection was completed. Certainly, the designer must have taken into account the behavior of this structure during the construction phase. Such items as stability, temporary bracing, and shoring must not be forgotten.

Finally, special loadings due to impact, uplift, and even temperature change must be a source of concern in design. Impact refers to the increased stress that a structure or member feels due to a sudden and dynamic load. Impact loading from vehicles or equipment may increase static stresses on members from 20% to 100%. Uplift loads from water and wind might be a concern in certain structures subject to unusual wind loads or structures located in floodplains. Temperature stresses caused by improper expansion joints or even temperature variations in members can cause unusual problems for the designer.[7]

The complete design will address all potential loadings in a manner consistent with a particular geographic location. This factor is a vital necessity for a safe and prudent design.

EXERCISES

1. Explain the differences between a building code and a specification. How are they similar?
2. Read the foreword of the AISC *Manual of Steel Construction,* and list the objectives of this organization. How do they differ from the objectives of a building code?
3. Review your state building code's chapter on structural loads, and list the values of ground snow load, basic wind speed, and seismic zone in regard to your school's geographic location. If you were designing an identical structure, how might you modify these values with respect to the importance factor, exposure factor, and other possible modifiers?
4. Again, review your state building code, and determine the applicable live loadings for your school. Is there more than one live load that may be applied in this structure?
5. For the beam shown here, calculate the weight per foot applied to the beam, using the information regarding dead and live loads given earlier in this chapter.

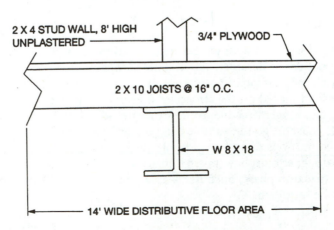

2 X 4 STUD WALL, 8' HIGH UNPLASTERED

3/4" PLYWOOD

2 X 10 JOISTS @ 16" O.C.

W 8 X 18

14' WIDE DISTRIBUTIVE FLOOR AREA

REFERENCES

1. Building Officials and Code Administrators International, *The BOCA National Building Code,* Country Club Hills, Illinois, 1990.
2. International Conference of Building Officials, *Uniform Building Code,* Whittier, California, 1991.
3. Southern Building Code Congress International, *Standard Building Code,* Birmingham, Alabama, 1991.
4. American Society of Civil Engineers, *ASCE Standard Minimum Design Loads for Buildings and Other Structures,* ASCE 7–88 (formerly ANSI A58.1), New York, 1990.
5. David P. Billington and John F. Abel, "Design of Cooling Towers for Wind," *Methods of Structural Analysis,* Proceedings of the National Structural Engineering Conference, Vol. 1 (New York: ASCE, 1976), p. 242.
6. Glen V. Berg, "Historical Review of Earthquakes, Damages, and Building Codes," *Methods of Structural Analysis, Proceedings of the National Structural Engineering Conference,* Vol. 1 (New York: ASCE, 1976), p. 387.
7. Mark Fintel and S.K. Ghosh, "Distress Due to Sun Camber in a Long-Span Roof of a Parking Garage," *Concrete International*, July 1988, pp. 42–50.

3

THE DESIGN AND PROBLEM-SOLVING PROCESSES

3.1 The Design Model

In the simplest sense, structural design is the process of solving a design problem. In this process the designer determines how resources, such as available wide-flange shapes, are to be allocated in a structural framing system. In this process, the designer will utilize his or her knowledge of engineering principles and materials to arrive at the optimum solution. An optimum solution will be economical and safe, yet take into account the individual needs that are specific to a particular project. But how does this process really take shape?

A variety of problem-solving concepts exist in many different fields of learning, such as business, information science, and psychology. Many people have developed these concepts into "design models" that can be applied universally in theory to any particular problem. Structural design also has a number of models that try to explain the steps taken by the structural designer to solve a particular problem.

Some design models portray problem solving as a spiral process,[1] in which the problem solver utilizes certain tools that stem from a large body of general knowledge and focuses down to a solution needed to solve a specific problem (Figure 3–1).

Other design models view problem solving in "input–output" terms,[2] where-

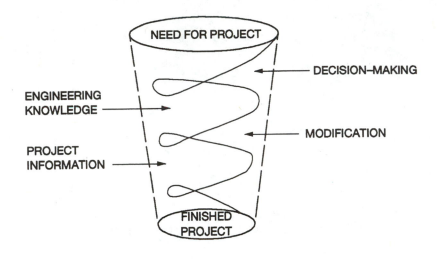

(PLANS, SPECIFICATIONS, COMPUTATIONS)

Figure 3–1   Spiral Design and Problem-Solving Process.

by certain information containing the specific requirements of the problem is input and, by utilizing various methods, the solution comes forth as the output. In this solution process, there is often reference made to the designer's "black-box" thought processes (Figure 3–2). The description of such thought processes as a black-box is meant to signify the individuality associated with design solutions because of the creative nature of this process. The inner workings and steps of any design will vary with the individual designer.

The events that occur inside the black-box are difficult to explain since they are unique to the structural designer. There are common elements that occur in each design process, and we will attempt to highlight some of them in the following paragraphs.

Every design process starts with **problem identification,** where the designer must "focus in" on all the design aspects with which he or she will be involved. For example, a column design will not involve merely the actual column design but also baseplate design, beam connection design, and connection of other appurtenances. To exclude these other facets of column design would not only be unwise, but also be potentially disastrous. A good design will always focus on the "big picture."

A design process will involve **trial-and-error techniques;** in fact, many times the most economical member is found by an iterative, trial-and-error approach. Although a trial-and-error approach may seem to be rather nontechnical, many famous structures from ancient times were actually "refinements" of similar

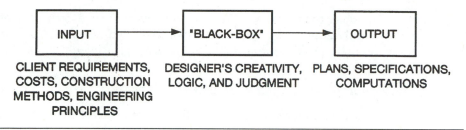

Figure 3–2   Black-Box Design Process.

structures that had succeeded or failed. A good historical example of this trial-and-error approach is the Bent Pyramid at Dashur. This pyramid, located in Egypt, is well known because at approximately midheight the angle of its side slope changes from 52° to 43.5°. This angle change is extremely unusual to have occurred during the construction of such a structure and has been attributed to the apparent failure of the nearby Meidum Pyramid.[3] It seems that when the news of the failure at the Meidum Pyramid reached the builders of the Bent Pyramid, they opted to reduce the slope in order to achieve (what they thought would be) a more stable design. Although many pyramids had been previously built using the 52° side angle, this one failure has been shown to influence later designs by what amounts to a trial-and-error approach. A good design will utilize trial-and-error techniques, not only in refinements of particular components, but also by way of previous engineering experiences. Many times we, as an engineering community, can learn more from our mistakes than from our successes.

Finally, all design processes will involve **computations,** which are a set of clearly written and precise design calculations. Computations are more than just formulas we use in the determination of certain structural components. Computations serve as our record of the design process and should include all of our assumptions, specification references, and justifications made during the design process. These computations are as close to a description of the events which occur inside the designer's black-box as there can be. Computations are immensely important because they serve as a designer's historical reference. This reference is important because it provides the designer with the assumptions made and the design steps utilized while working on a specific type of project. It may also become very important should the designer become involved in litigation.

Although preparing a good set of computations is vitally important, this talent is usually one of the hardest for a young designer to acquire. As a general rule, the author believes that if a designer can stop working on a particular set of design computations for two weeks and then resume work with a minimum amount of time for startup, then the designer has a relatively good set of design computations.

In closing, the aforementioned concepts are only some of those involved in a structural design process. Many other aspects will vary, depending on the individual designer and the type of project. However, the idea of structural design as a problem-solving process is extremely important because it reinforces that design is based on logic, creativity, and sound engineering principles.

3.2   The Assumption of Load Path and Floor Loading

One of the first assumptions a structural designer makes is the path across which the forces travel as they move throughout a structure. Loads (or forces) will travel through a structure along what is referred to as a *load path*. The location of the assumed load path can be found conveniently by answering the question, "How do the loads on the structure get transferred to the supporting soil?"

Structures may be viewed as mechanisms by which loads are distributed to their individual members, such as beams, columns, and floor slabs. It remains up to the structural designer to make judgments on the amount of load distributed to individual members and the manner in which loads will travel throughout the structure.

Many times a structural designer will assume a distribution of loads to particular members based on the concept of *distributive area*. This concept typically considers the area that a member has to support as being halfway between the next closest, similar member(s). To illustrate, Figure 3–3 shows that the steel floor beam located at 48 in. on center is assumed to carry the dead and live loads placed in the area 24 in. to the left and 24 in. to the right of the beam.

The distributive area technique works very well in normal floor framing under standard dead and live loads. A typical steel framing plan for a supported floor system may be found in Figure 3–4. In this illustration we can see that the floor beams at 6 ft on center (o.c.) are tied into larger floor girders. These girders are then connected to the columns located at each of the four corners of the bay. (A *bay* typically refers to the area bounded by structural columns, usually in a square or rectangular arrangement. In Figure 3–4, one bay is bounded by columns A1–A2–

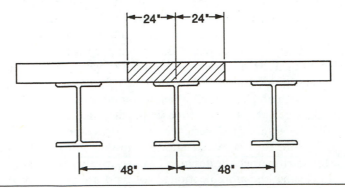

Figure 3–3   Distributive Area in a Structural Floor Slab.

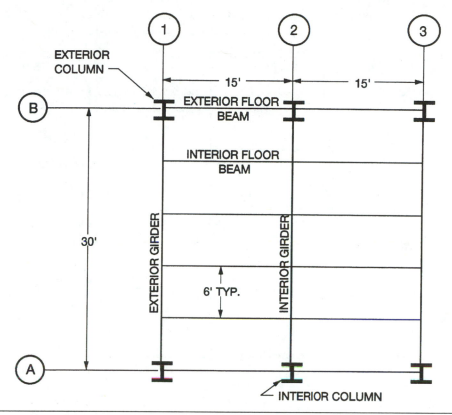

Figure 3–4   Typical Floor Framing System.

B2–B1 and the other bay is bounded by columns A2–A3–B3–B2.) Under the distributive area technique, the designer will assume that each of the floor beams carries the loads in its area and that these loads are then transferred as concentrated loads to the larger girders, which in turn distribute their loads to the columns.

The following example will illustrate how the distributive area technique may be used to approximate not only the amount of load on individual members but also the load path that exists.

EXAMPLE 3.1

Calculate the loads on each member (floor beams, girders, and columns) shown in the framing plan in the accompanying figure, using the distributive area method. The floor live load is 100 psf, and the dead load is 120 psf (which includes slab and member self-weight).

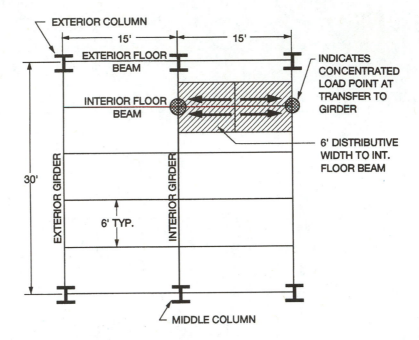

Initially, the designer must realize that there is a big difference in quantity of load applied to the interior (or center) beams, girders, and columns as opposed to the exterior (or edge) beams, girders, and columns. This difference is attributed to a larger amount of distributive area for these interior members.

For the Interior Floor beams (spaced at 6 ft o.c.)

Distributive area is 6 ft; therefore each interior beam feels the application of 1320 lb per lineal foot (or 6 ft × 220 psf). If each beam is assumed to be simply supported over its 15-ft length, the reaction placed on the floor girder from each interior floor beam is as follows:

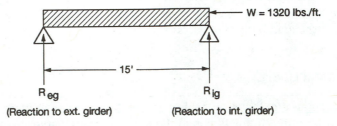

$$R_{ig} = R_{eg} = wl/2 = (1320 \text{ lb/ft})(15 \text{ ft})/2 = 9900 \text{ lb}$$

For the Exterior Floor Beams (which frame directly into the columns)

Distributive area is 3 ft; therefore each exterior beam feels the application of 660 lb per lineal foot (3 ft × 220 psf) over its 15-ft length. Assuming the simple support conditions, the reactions applied directly to the columns would be as follows:

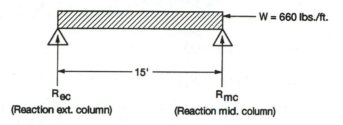

$$R_{ec} = R_{mc} = wl/2 = (660 \text{ lb/ft})(15 \text{ ft})/2 = 4950 \text{ lb}$$

For the Interior Floor Girder

The reaction of 9900 lb from the floor beams is applied every 6 ft in the locations shown in the following diagram. Remember that for the interior girder the floor beams frame in from each side, so the total applied load at the four points shown is doubled. These loads are then transferred to the middle column.

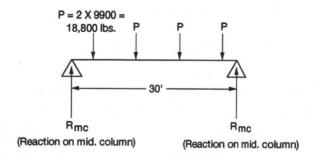

$$R_{mc} = 4P/2 = 4(19,800 \text{ lb})/2 = 39,600 \text{ lb}$$

For the Exterior Floor Girder

The reaction of 9900 lb from the floor beam frames in from one side only; therefore, the load is only 9900 lb at each location. As shown in the following diagram, the reaction at the exterior columns would therefore only be half as much.

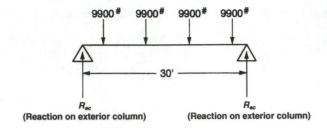

$$R_{ec} = 4(P)/2 = 4(9900 \text{ lb})/2 = 19{,}800 \text{ lb}$$

For the Middle Columns (R_{mc})

The reaction applied to the column from the two exterior floor beams (each side of column) and the interior girder reaction results in a total load as follows:

$$P_{mc} = 2(R_{mc})_{\text{ext. fl. beams}} + (R_{mc})_{\text{int. gird.}}$$
$$= 2(4950 \text{ lb}) + (39{,}600 \text{ lb}) = 49{,}500 \text{ lb}$$

For the Exterior Columns

The reaction applied to the column from the one exterior floor beam and the exterior girder is as follows:

$$P_{ec} = (R_{ec})_{\text{ext. fl. beams}} + (R_{ec})_{\text{ext. gird.}}$$
$$= (4950 \text{ lb}) + (19{,}800 \text{ lb}) = 24{,}750 \text{ lb}$$

3.3   The Process of Member Evaluation

Since we have discussed (at some length) the concept of problem-solving processes, let us begin to focus our attention on the particular problems that occur in structural design. Only two general categories of structural design problems exist in the realm of engineering: **evaluation problems** and **design problems.** If the inexperienced designer can establish the type of problem with which he or she is dealing, the solution process can be clearly defined. In this section, we will discuss the problem-solving process contained in member evaluation problems; the following section will present the solution process in design problems.

Evaluation problems require the structural designer to evaluate an existing member under existing or proposed loading. Evaluation problems may ask the designer to answer one of the following questions:

- Is the member adequate under the specified load?
- Will the member meet specification criteria?
- Is the member safe for public usage?

When using the Allowable Stress Design method, all these questions can be grouped into one general question that lies at the heart of the ASD philosophy: Is the member's actual stress less than or equal to its allowable stress? If the designer's answer to this question is yes, the member is adequate and the evaluation problem for that behavior is ended. Should the answer to this question be no, the designer must initiate design changes to correct the deficiency.

The evaluation problem always has an existing member with specific dimensions and material properties. If fortunate, the designer will have access to an existing set of plans or specifications that will readily show the member's dimensions and properties. However, on many older structures, a set of plans or specifications may no longer exist. Therefore, a detailed inspection and testing will have to be performed to attain the member size and material properties.

In addition to knowing (or being able to attain) the member size and properties, in an evaluation problem the designer always knows the load or loadings for use in evaluating the adequacy of the member. Knowing this information, the designer can always calculate the actual stress that will be distributed to the member. The two most common stress equations (which the student will remember from strength of materials) are the direct stress formula and the bending stress formula:

$f = P/A$ (direct stress)

$f = Mc/I$ (bending stress)

where f = actual stress.

The direct stress formula is used many times throughout structural design to calculate such items as tensile stress, compressive stress, and shear stress. The bending stress formula is used to calculate bending stress (of course) and combined stress, where bending stresses may be coupled with other types of stress.

After calculating the actual stress, the designer will then use the knowledge gleaned of the particular problem and the specification being used to calculate the allowable stress, F. (It should be noted that in steel design the actual stress is referred to by the lowercase letter f, while the allowable stress will be referred to by the uppercase letter F.) The comparison of actual stress to allowable stress can then be readily accomplished and will be used extensively in the upcoming chapters.

3.4   The Process of Member Design

Design problems are usually considered to be the harder of the two types facing the structural designer. This is because the structural designer is required to choose a particular section that will provide sufficient strength and serviceability and fit into the architectural system for which it is designed. It is much easier to consider the adequacy of a given member (with all of its dimensions and properties known) than to be given general requirements and asked to design.

The problem-solving process of design problems can be somewhat simplified if a designer can focus attention on what really is to be solved. In general, the solution to all design problems is finding the most suitable (and economical) member area. This member area may actually be expressed as an area (i.e., square inches) or as some function of area (i.e., moment of inertia, in.^{4}, or section modulus, in.^{3}). The most suitable area will be that which provides an actual stress somewhere close (if not equal) to the allowable stress level. If the designer chooses an actual member size that provides an actual stress equal to the allowable stress, the member is (in theory) as economical as it can be without exceeding the specified allowable stress.

Therefore, the two most commonly used design formulas are still the direct stress equation and the bending stress equation. However, the designer is now solving for an area (or a function of area), using as the stress level the maximum allowed per specification (the allowable stress). The reworked direct stress and bending formulas used in design are now as shown below:

$$A_{min} = P/F \qquad \text{(direct stress)}$$

$$(I/c)_{min} = M/F \quad \text{(bending stress)}$$

where

F = the allowable stress level to be used (these will usually have a subscript attached to identify the type of behavior; for example, the allowable bending stress will be referred to as F_b)

A_{min} = minimum area that will work

$(I/c)_{min}$ = minimum section modulus that will work

(*Note:* The ratio I/c is sometimes designated as the section modulus, S. More will be discussed about this in Chapter 6.)

The student will find that the design process involves many choices and is usually very challenging. Design should be viewed as a continuous problem, one that will not be solved on the very first try. In fact, many trials (and errors) are usually needed to refine the problem into its final form. In many ways, the structural design process is similar to the process in which music might be composed. Although there are only a few notes on the musical scale (and only a few types of structural members), there can be an infinite number of ways to put them together.

3.5   The Importance of Design Details

The construction of any structure, from a house to an 80-story skyscraper, involves a seemingly infinite number of "small" items that require attention. These "small" items are typically referred to as details, although they are not in any way insignificant.

Design details are not the major structural components (such as beams and columns) that we typically think of when we consider the process of design. Rather, design details may focus on more subtle design items such as connection details between members, expansion and contraction devices, stairway and handrail framing, and mechanical system support. Design details are numerous and may literally outnumber the design of the "major" structural components by a wide margin.

To a structural designer, design details may sometimes seem unglamorous. For instance, when a bridge is designed, how many people notice the expansion joint system as opposed to the main supporting trusses? Although the general public may never notice the expansion joint, its proper design and correct detailing is probably every bit as important to maintaining the bridge's longevity. Every design detail, although time-consuming and tedious, must be fully considered and correctly designed to ensure the safety and serviceability of every structure.

One of the most catastrophic examples involving the failure of a design detail was the collapse of two suspended walkways in the Hyatt Regency Hotel in Kansas City, Missouri, on July 17, 1981 (Figure 3–5). This disaster, which killed 114 people and injured more than 200, resulted from a failure of the box beam–hanger rod connection (Figure 3–6).

The collapse occurred during an early evening dance being held in the hotel's atrium and was believed to have initially occurred in the middle box beam on the fourth floor walkway. The collapse mechanism was probably initiated when the nut assembly of the hanger rod pulled through the box beam, leading to similar failures at the other fourth floor connection points. As the fourth floor walkway lost support, it collapsed on the second floor walkway below, resulting in "the most devastating structural collapse ever to take place in the United States."[4]

Although much discussion following the collapse involved a change that was made to the original hanger rod detail, the following important points can be made:

- The loads acting at the time of the collapse were only 53% of those required by the Kansas City Building Code. Neither the original connection design nor the revised detail would have supported the specified loads.
- An investigation by the Deutsch Commission (which was to rule on a complaint filed by the Missouri State Board to revoke the licenses of the engineers involved) revealed that no original design calculations regarding this connection were found.
- Communication failure between the structural designer and the steel fabricator regarding the responsibility for the failed connection design was a major contributing factor in the collapse.[5]

How did this design detail slip between the cracks? Was it lost among the "more important" structural design items? Although we cannot attempt to answer these

Figure 3–5   Scattered Wreckage in the Aftermath of the Kansas City Hyatt-Regency Collapse. (Courtesy of World Wide Photos.)

questions, this collapse underscores the importance of all pieces in the puzzle that we refer to as *design*.

3.6   Computers in Structural Design

Computers, and the software that accompanies them, have invaded practically every facet of our business and personal lives. Structural engineering and design is no exception. As early as 1958 the American Society of Civil Engineers convened a conference to discuss the use and potential of computers in structural engineering. Although the original computers were huge, relatively slow machines requiring extensive setup time by the user, these machines were the pioneers of the revolution that we commonly refer to as the "computer age."

Advances and improvements in hardware and software have been especially dramatic in the last decade. Hardware advances have led to the use of personal computers by everyone in the structural design field. Personal computers have

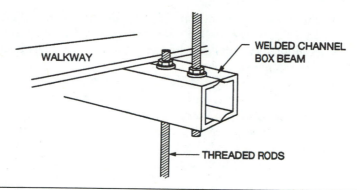

Figure 3-6   Schematic Illustration of Box Beam–Threaded Rod Connection Involved in Walkway Collapse.

become more affordable, while also becoming faster and more powerful in their ability to assimilate data. Software advances have also accelerated in the last ten years. Many structural analysis programs provide the user with interactive design modules that essentially prompt the user with specific questions concerning the design. Computer-aided drafting (CAD) software has rapidly advanced to the point that many analysis software packages now can produce framing plans and other structural drawings. Other software packages on the market contain typical structural details known as "library symbols," which can be modified to suit a particular design and transferred or "imported" via transfer files into the actual CAD software being used.

The software industry has begun to refine this transfer mechanism (using DXF, IGES, or DXB files) and will continue to do so for the future. The ability to transfer files from one type of software to another is an extremely powerful tool because it allows greater flexibility in the design process. A designer can now draw a "wire-frame model" of a structure to be analyzed in the earliest stages of preliminary design and continue to refine and modify this model until the design is complete. The designer then can use this model as the basis for the structural plans and, through the use of other detailing software, produce a complete set of structural plans.

Although the use of computer technology in structural design is expanding, it is important for the student to remember the following:

- The computer and its software are only tools for the designer's use. Although some may think that a computer can design, in actuality, it can only process data. Always check the computer's results. Although some people would like to believe that computers are infallible, in structural design, this assumption would be potentially disastrous.

- The most important function of a designer is the ability to communicate and justify the design that has been selected. Although some of the computer's results may be used in a designer's justification, the computer alone is not enough.
- Do not be intimidated by computers or their software. The only way to learn about a piece of software is to spend many hours examining (or "playing with") it. Learn about all different types of software, from design and analysis, to CAD software, and even to word-processing software. At first, some of this "play" time may not seem to have much benefit; but in the end it will increase your marketability and expertise.

EXERCISES

1. The cross section shown in the accompanying figure is the framing for a 70-foot-long simply supported bridge. Using the distributive area method, calculate the dead load (per foot) carried by an interior and an exterior girder. Assume that the parapet and railing will be distributed equally among all girders. The areas and densities of the materials are as follows:

   $\gamma_{concrete}$ = 150 pcf

   Area of parapet = 3 ft^2

   Area of beam = 7.8 ft^2

   1-in.-thick future wearing surface, γ = 120 pcf

   Guardrail = 15 lb/ft

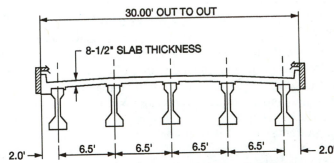

2. Investigate your house or apartment's framing system. Roughly sketch the floor plan, and decide how loads travel throughout this structure from the roof to the foundation.
3. Give another specific example of trial-and-error design of some type of historical structure. (*Hint:* Gothic cathedrals, early railroad bridges, the Parthenon of ancient Greece.)

REFERENCES

1. James M. Becker, "A Structural Design Process," *Proceedings of the National Structural Engineering Conference, Methods of Structural Analysis,* Vol. 2, ASCE, 1976, p. 341.

2. William Addis, *Structural Engineering, The Nature of Theory and Design* (London: Ellis Horwood Limited, 1990), p. 37.

3. Mario Salvadori, *Why Buildings Stand Up* (New York: W.W. Norton and Company, 1980), p. 38.

4. E.O. Pfrang and R.M. Marshall, "Collapse of the Kansas City Hyatt Regency Walkways," *Civil Engineering,* July 1982, pp. 65–68.

5. E.A. Banset and G.M. Parsons, "Communications Failure in Hyatt Regency Disaster," *Journal of Professional Issues in Engineering,* July 1989, pp. 273–288.

4

TENSION MEMBER DESIGN

4.1 Introduction to Tension Members

The use of steel tension members is found throughout building and bridge construction today. The bottom chord members of simply supported bridge and roof trusses, as well as the cables used in a variety of suspended bridges and walkways, are all tension members. (See Figure 4–1.)

Tension members may be cables, rods, angles, channels, wide-flange sections, plates, or any number of sections that are built up from these individual shapes. When builtup sections are used, they may be held together by tie bars or lacing bars located at certain intervals along the member's length to hold the individual pieces in correct alignment. The type of tension member used is largely a function of its end connections. A cable or circular rod might be an ideal tension member, although its physical shape may require special detailing in its connection to other members.

4.2 Modes of Failure and Formula Usage

The design of tension members in this book will follow the Allowable Stress philosophy, as previously mentioned. This method requires that we keep the stresses in our member below some acceptable limit set forth by the AISC specification. The direct stress formula for calculating axial stress will be our most useful equa-

Figure 4–1   Example of Cables as Tension Members in the George Washington Suspension Bridge. (Courtesy of Bethlehem Steel Corporation.)

tion. The student will soon realize that every problem in tension member design uses some form of the direct stress formula. This equation, which is the cornerstone of the curriculum in strength of materials, is $f = P/A$, where

f = actual stress, psi or ksi

P = load, pounds, or kips

A = resisting area, square inches

Before learning how to apply this equation in tension member design, it is necessary to consider the common modes of failure that might occur when a member is stressed in tension. These modes can be summarized as follows:

- The member fails where there are no holes.
- The member fails where there are holes for bolted or riveted connections.
- The member fails due to inadequate bolt spacing.

The third method of failure should be the least likely to occur, since the AISC specification lists (in Chapter J) typical maximum and minimum requirements for distances between bolts and edge spacings. A good engineer or structural detailer will know these requirements, and the student will further touch upon these specific criteria in Chapter 9 of this book. At this time, the interested student is referred to Section J3 of the AISC *Manual for Steel Construction,* where these requirements are located. Other modes of failure sometimes associated with tension members, such as bolt shear and bearing, are considered by this author to be more closely identified with connection failure and are therefore the subject of Chapter 9.

The first two methods of failure in the preceding list truly constitute the basis of what the designer is trying to prevent in tension member design. Failure through the member in a location where there are no holes is typically referred to as failure in the gross area, while failure through the plate in the area of bolt holes is referred to as failure in the net area. The gross and net areas are radically different in their individual behaviors and their modes of failure. Therefore, it is imperative that both of these areas must be understood fully. In *every* case (excluding rods), both the gross and net areas must be checked to determine which of the two is most critical.

As mentioned earlier in this section, the direct stress equation can be used in a number of ways either to determine a member's adequacy per code requirements or to design the member. In design, the direct stress formula can be used to determine the minimum required cross-sectional area for the member (A_{min}). This is the smallest area (hence the most economical) that would meet the allowable stress criteria of the AISC specification. The formula for the minimum required area is simply a rearrangement of the direct stress formula. This rearrangement can be shown as follows:

$$f_t \leq F_t$$

$$P/A \leq F_t$$

$$P \leq F_t \times A$$

$$P/F_t \leq A$$

Therefore,

$$A_{min} = P/F_t$$

where P = load on member to be designed and F_t = allowable stress (per AISC specification).

To evaluate an existing member, the direct stress formula can be used in two different ways to determine the member's adequacy. The allowable load that a member can hold (P_{all}) can be calculated and compared to the actual load on the member. If the allowable load is greater than the actual load, the member is indeed adequate. The other method to evaluate a member compares the actual stress on the member (by using the direct stress formula) to the allowable stress level as given by the specification. If the actual stress on a member is less than or equal to the allowable stress as given by the specification, again the member is adequate.

These methods of investigating the member's capacity by the direct stress formula are shown in the following two formulas:

- $P_{all} = F_t \times A$; if $P_{all} \geq P_{actual}$, then member is okay.
- $f_t = P_{actual}/A$; if $f_t \leq F_t$, then member is okay.

where

A = known cross-sectional area of a member

A_{min} = minimum cross-sectional area of a member to be designed

P_{actual} = known tensile load

P_{all} = maximum allowable tensile load

F_t = allowable tensile stress (per AISC specification)

f_t = actual tensile stress

Now that the student understands the methods behind the formula usage for tension members, the following section will be devoted to the understanding of the anticipated behavior of tension members, as well as to the solution of common problems.

4.3 Tension Member Behavior: Gross Versus Net Area

The gross area is the cross-sectional area of the tension member that has no bolt or rivet holes (Figure 4–2). This is the area away from the connection that is resisting the tensile stress. The actual stress in this location is very well approximated by the standard direct stress equation, $f_t = P/A$. The failure mechanism of tension members across the gross area is relatively simple: It would occur due to the yielding of the steel. Since the stress over the gross area is relatively uniform, yielding of all individual fibers of the cross section is anticipated to occur at the same time. Therefore, the failure we are trying to prevent, with regard to the gross area, is a yielding type of failure. The student can imagine this failure as occurring by the member's stretching out in an uncontrollable fashion, never able to regain its original length. (Remember the stress–strain diagram?)

The AISC specification (in section D1) sets the allowable tensile stress over the gross area for rolled members as $0.60 \, F_y$. That is, the specification is setting the maximum value for tension stress over the gross area (which designers should not exceed) at 60% of the steel's yield stress. For A36 steel (with a yield stress, $F_y = 36$ ksi) this value would be 0.60×36 ksi or 21.6 ksi (some designers would round this value to 22 ksi).

Stresses on a tension member over the net area (i.e., vicinity of bolt holes) are extremely complex, and studies in experimental stress analysis have shown that stresses immediately adjacent to holes may be 2 to 4 times greater than those across the gross section.[1] To produce failure over this section, it must be remem-

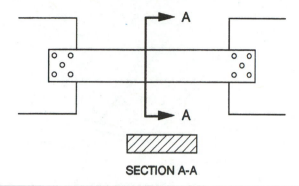

SECTION A-A

Figure 4–2   Illustration of Gross Area in a Tension Member.

bered, the full cross section must fail by the same mechanism. Although the full gross area failed by reaching its yield stress, the net area has stresses that are very uneven over its cross section. Since the areas directly adjacent to the bolt holes are much more highly stressed, these areas begin to yield long before the other areas of the net cross section. However, since much of the cross-sectional area is not even close to reaching its yield stress, the section will not fail by yielding at this time.

As stresses continue to build in the net area, the locations that have reached yield stress initially (the areas adjacent to the bolt holes) are now approaching the steel's ultimate tensile strength, F_u. At this time, cracking occurs at the holes and quickly propagates outward, resulting in a fracture of the member (Figure 4–3).

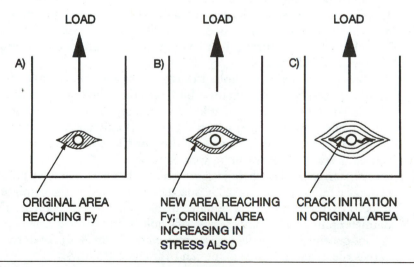

Figure 4–3   Failure Propagation Around Bolt Holes.

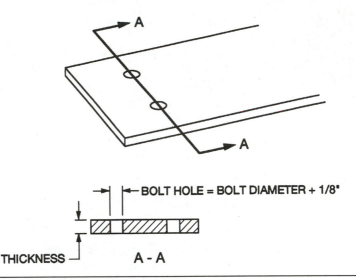

Figure 4–4   Illustration of Net Area in a Tension Member.

Some redistribution of stresses does occur, but the mode of failure across the net area is a fracture type of failure. Therefore, in design, we try to prevent this cracking failure from occurring at the net area.

Because of this behavior, the AISC specification sets the allowable tensile stress across net areas as $0.50 F_u$, or 50% of the steel's ultimate tensile strength.

The calculation of net area also has more factors to consider than does gross area. In the easiest sense, the net section is the gross area minus the area for the bolt holes in the section considered. Actually, there may be two or three net areas to be considered when a multiple-bolt layout is used. Multiple-bolt configurations will be discussed in Section 4.5.

Before subtracting the holes from the gross area, the designer must remember that bolt holes have to be made a bit larger than the bolts so that erection is not hampered. Generally, the method for making holes in steel is a process of subpunching and reaming, which causes little damage to the sides of the holes. This process is typically faster[2] and more cost effective than drilling. If this process is used, a standard hole will be approximately 1/16 in. larger than the bolt.

The area of a bolt hole is considered to be its rectangular projection on the proposed failure plane. The AISC specification considers the width of the bolt hole to be 1/16 in. larger than the nominal diameter of the hole, which is simply the bolt diameter plus 1/8 in. Thus, the area of a hole is simply the diameter of the bolt plus 1/8 in. multiplied by the thickness of the plate (Figure 4–4). Therefore, the net area across the assumed failure plane can be found as follows:

Net area = Gross area − [(number of bolts) ×
(diameter of bolt + 1/8 in.) × (plate thickness)]

The following examples will introduce the student to the methodology used in solving tension member problems. *Remember:* It is imperative to check both the gross and net areas in every problem.

EXAMPLE 4.1

The tension member shown in the figure is subjected to an 85-kip load. Calculate the actual stresses across the gross and net areas, and check their adequacy per AISC requirements. The steel is A36, and the bolts are 3/4 in. diameter.

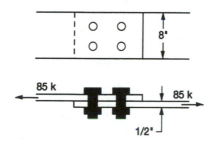

(In this problem we are asked to evaluate a given member under a given load. Remember that there are two methods of evaluation, and in this example we will look at the comparison of actual stress to the AISC values of allowable stress.)

The tension member is a plate; therefore, calculate the gross and net areas as follows:

A_{gross} = 8 in. × 1/2 in. = 4 in.^{2}

A_{net} = 4 in.^{2} − 2(3/4 in. + 1/8 in.)(1/2 in.) = 3.125 in.^{2}

Now calculate the stress across the gross area by the direct stress formula and compare your result to the specified value of allowable stress:

f_t (actual) = 85 kips/4 in.^{2} = 21.25 ksi

AISC specification allows 0.60 F_y = 21.6 ksi. Since actual stress is less than allowable stress, the member is okay along gross area.

Likewise, calculate the stress across the net area and also compare to the specified value of allowable stress:

f_t (actual) = 85 kips/3.125 in.^{2} = 27.2 ksi

AISC specification allows 0.50 F_u = 29 ksi. Since actual stress is less than allowable stress, the member is okay across net area.

Since the member is adequate at both gross and net areas, the member will work per AISC specification.

EXAMPLE 4.2

Design the thickness of the flat plate tension member shown in the accompanying figure, if the load needed to be supported is 55 kips. The bolt is 3/4-in. diameter, and the steel is A36.

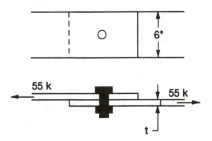

(This is a design problem, since thickness of a plate is a function of its area. Remember to consider both gross and net areas.)

Gross area: A_{min} = 55 kips/F_t

Where F_t, the allowable tensile stress per AISC, = 0.60 F_y, or 21.6 ksi. Remember that the smallest area that still meets AISC criteria is the most economical and therefore is the best design. The stress level that will give us this smallest area and still meet AISC criteria is the allowable stress level. Therefore,

A_{min} = 55 kips/21.6 ksi = 2.55 in.^{2}

Since thickness = area/width = 2.55 in.^{2}/6 in.

Thickness = .43 in.   or   7/16 in.

Now consider net area:

Net area: A_{min} = 55 kips/F_t

where F_t is the allowable tensile stress per AISC = 0.50 F_u or for A36 steel, F_t = 29 ksi

A_{min} = 55 kips/29 ksi = 1.90 in.^{2}

1.90 in.^{2} = minimum net area that will work

Express net area in terms of thickness, t, as follows:

Net area = $6t - (1)(.75$ in. $+ 1/8$ in.$)(t) = 5.125t$

1.90 in.^{2} = $5.125t$

.371 in. or 3/8 in. = t

We have calculated two thicknesses (one for the gross area and one for the net area), but the critical thickness would be 7/16 in. (gross area controls). This is because, although a 3/8-in.-thick plate would work for the net area, it would be too small for the gross area.

(*Author's note:* In striving to present a simplified example to the beginning designer, I have illustrated a tension member design using a connection of only one bolt. This connection would undoubtedly fail in bolt shear and provide no structural redundancy in case of material deficiency or other potential failure catalysts.)

4.4   Effective Net Area, A_e

The student may have noticed that, up to now, all the illustrations and examples on tension members have dealt only with flat plates. This was purposely done in the initial discussions to simplify the subject of net area. Flat plates are ideal tension members with regard to their net area because all of their net area lies in the plane of loading. Therefore, all of a flat plate's net area is assumed to be active in resisting the tensile stress. This is not the case with rolled shapes, whose full cross section does not lie in the plane of the loading.

Previously, in the discussion on net area, it was revealed that the net section has an extremely uneven stress distribution. The magnitude of this stress distribution is increased dramatically if all of the net area is not effective in contributing to the resistance of tension stress. To measure a rolled shape's effectiveness of net area, the AISC introduces a term, A_e, which is called the effective net area. The effective net area, A_e, is simply a member's net area multiplied by a reduction coefficient, U. This is shown as follows:

$A_e = A_{net} \times U$

This reduction factor, U, approximates the increase in stress near the connection due to shear stress concentration when the full net area is not effective in transferring

Table 4–1   Standard AISC Reduction Coefficients, U

Description	Requirements		U
1. W, M, S shapes connected to flanges, and structural tees cut from these shapes	$b_f \geq (2/3)d$	3 bolts per line minimum	.90
2. W, M, S shapes and builtup members not meeting 1, and all other members meeting the requirements		3 bolts per line minimum	.85
3. All members having only 2 bolts per line			.75

Note: Values to be used unless larger values can be justified by testing.

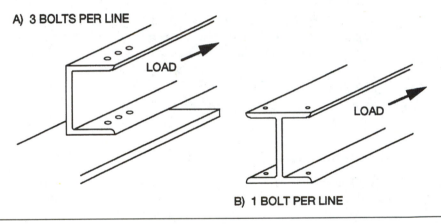

A) 3 BOLTS PER LINE

B) 1 BOLT PER LINE

Figure 4–5   Illustration of "Bolts per Line."

the applied load. The approximation of increased stress is actually done by decreasing the amount of area available. Hence, the term *effective net area.*

The value of U can most easily be found in Section B3 of the AISC specification or taken from Table 4–1. These values reflect an approximate estimate of effectiveness in a rolled section based on items such as the type of section, the stockiness of the section, and the length of connection. The only elaboration needed on the information found in Table 4–1 is the explanation of the number of bolts per "line." A line of bolts refers to the fasteners spaced along the length of the member, *not* across the net section (Figure 4–5). The more bolts there are per line, the longer is the connection area, thus helping to create a more effective net area.

The direct stress formula for calculating either the minimum required cross-sectional area, the maximum allowable load, or the actual stress at the effective net section now becomes:

$A_{\text{net min}} = P/(U \times F_t)$

$P_{\text{all}} = F_t \times A_e$

$f_t = P/(A_e)$

where

$A_{\text{net min}}$ = minimum net cross-sectional area of member to be designed

A_e = effective net area (UA_n)

P = known tensile load

P_{all} = maximum allowable load

F_t = allowable tensile stress at effective net area $(0.50\ F_u)$

f_t = actual tensile stress at the effective net area

The following examples will illustrate the use of this reduction coefficient, U, as it relates to the effective net area.

EXAMPLE 4.3

The angle shown in the accompanying figure is a 5 × 5 × 1/2 and is under a tensile load of 48 kips. The 7/8-in.-diameter bolts are connected to the gusset plate, as shown, and all steel is A36. Calculate the adequacy of this member per AISC requirements.

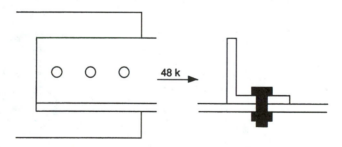

(Again, this is an evaluation problem to check the adequacy of a known member under a given load. This time we will check adequacy by comparing allowable load to the actual load of 48 kips.)

$A_{\text{gross}} = 4.75$ in.^{2} (See shape table in Appendix B or AISC manual, pp. 1–47.)

$P_{\text{all}} = F_t \times A_{\text{gross}}$

Since the allowable stress over the gross area is 0.60 F_y, P_{all} is calculated as follows:

$$P_{all} = 21.6 \text{ ksi} \times 4.75 \text{ in.}^2 = 102.6 \text{ kips}$$

which is greater than the 48 kips of actual load; therefore, the gross area is okay per AISC specification.

Check the effective net area:

$$A_e = UA_n$$

$$A_{net} = 4.75 \text{ in.}^2 - (1)(7/8 \text{ in.} + 1/8 \text{ in.})(1/2 \text{ in.}) = 4.25 \text{ in.}^2$$

$$U = .85 \text{ (not a W, M, or S shape, 3 fasteners per line)}$$

$$A_e = .85 \times 4.25 \text{ in.}^2 = 3.61 \text{ in.}^2$$

Since $F_t = 0.50 \ F_u = 29$ ksi,

$$P_{all} = 3.61 \text{ in.}^2 \times 29 \text{ ksi} = 104.76 \text{ kips}$$

which is greater than 48 kips. Therefore, effective net area is okay per AISC specification. Since both areas meet AISC criteria, member is adequate.

EXAMPLE 4.4

Design a W12 section to hold a tensile load of 150 kips. The connection is to be made by 3/4-in. bolts through the flanges, as shown in the figure. There are at least three bolts per line, and the steel is A36.

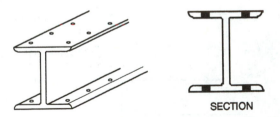

SECTION

(In this design problem, we are trying to determine the minimum area of a W 12 section that will work. Since there are at least three bolts per line, we should know that the reduction factor, U, is either .85 or .90, based on the flange width to depth criteria.)

To begin with, design for gross area, realizing that the allowable stress is 0.60 F_y, or 21.6 ksi (this stress level will give us the minimum workable area in regard to the gross area):

$$A_{min} = 150 \text{ kips}/21.6 \text{ ksi} = 6.95 \text{ in.}^2$$

Looking in the section tables at the smallest possible W12 to meet this criterion, we would try a W 12×26 ($A = 7.65$ in.^{2}).

Check the effective net area based on a W 12×26. The reduction factor for a W 12×26 (connected as shown) is .85, since 2/3 times depth (8.15 in.) exceeds the flange width (6.49 in.). Therefore, $A_{\text{net min}}$ = 150 kips/(29 ksi × .85) = 6.09 in.^{2} (where 29 ksi = 0.50 F_u).

Because we have already incorporated the reduction factor, U, in the preceding calculation, we can simply compare that net area to the actual net area of a W 12×26:

$$A_n = 7.65 \text{ in.}^2 - (4)(.75 \text{ in.} + 1/8 \text{ in.})(.38 \text{ in.}) = 6.32 \text{ in.}^2$$

Since our actual net area is greater than the minimum (6.08 in.^{2}), the W 12×26 will indeed be okay.

4.5 Multiple-Bolt Configurations

In the discussion of effective net area's failure mechanism, we have implicitly considered the fracture to occur only directly across the member. That is, we have considered the cracking to occur only perpendicular to the direction of stress. This cracking failure, however, does not always occur straight across the member, but rather will occur in the path of least resistance. This path of least resistance is the path traversing the smallest effective net area.

The bolt arrangement in Figure 4–6 has a layout consisting of multiple lines and spacings. Therefore, it is not readily apparent if failure would occur in the net area at section A–A or at section B–B. Although the failure line in section B–B is much longer, it also traverses two bolt holes instead of the one that section A–A would fail through. Which is the critical net area?

Although there are theoretical formulas that consider the shear stress effect along the diagonal failure line between bolt holes, tests show that little is gained from using these complicated formulas over the empirical formula that the AISC proposes.[3] This empirical formula merely takes the net area (as we previously calculated) and adds in the additional diagonal areas by using the term $S^2/4g$. In this expression, S is the longitudinal spacing of bolt holes (center to center) and g is the gauge distance, or the center-to-center bolt spacing across the width of the member. Students should think of the $S^2/4g$ term as the diagonal distance between bolt holes, although they should remember that it is empirical by nature (it cannot be derived strictly by mathematical theory).

Therefore, when calculating the net area across a section containing diagonal areas, the net area formula could be written:

$A_{\text{net}} = A_{\text{gross}} - [(\text{number of holes})(\text{diameter} + 1/8)(\text{thickness}) + (\text{number of diagonals})(S^2/4g)(\text{thickness})]$

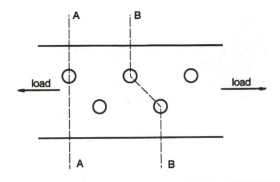

Figure 4–6   Potential Failure Paths with Multiple-Bolt Layouts.

Every plausible net area should be calculated to determine which is critical. It should be noted that the $S^2/4g$ term is an approximation that the AISC gives for plates and angles. When dealing with more complex shapes such as channels, wide flanges, and builtup members, the engineer's judgment should be used regarding the various possibilities of critical net area.

The following example will illustrate the determination and use of critical effective net area in a standard tension member problem.

EXAMPLE 4.5

Determine the adequacy of the plate shown in the accompanying figure, if it is made from A36 steel. Be sure to calculate the effective net area in the plate, considering sections through A–B, A–C–B, and A–C–E. The plate thickness is 1/2 in., and the bolt diameter is 5/8 in.

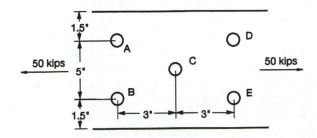

(In this evaluation problem, we will determine adequacy by calculating the allowable load and comparing it to the actual load of 50 kips. Since we are dealing with a flat plate, the reduction factor, U, for the effective net area is equal to 1.0.)

Consider gross area:

$P_{all} = A_g \times F_t$ (A_g = 8 in. × 1/2 in. = 4 in.^{2} and F_t = 0.60 F_y = 21.6 ksi for A36 steel)

P_{all} = 4 in.^{2} × 21.6 ksi = 86.4 kips

Since 86.4 kips is greater than the actual load of 50 kips, the member is okay across gross area.

Consider the effective net area (remember, since U = 1.0, effective net area is simply the net area). (In this problem, s = 3 in. and g = 2.5 in.; therefore, $s^2/4g$ = .90).

Net area: A–B: 4 in.^{2} – (2)(.75 in.)(.5 in.) = 3.25 in.^{2}

A–C–B: 4 in.^{2} – (3)(.75 in.)(.5 in.) + (2)(.90)(.5 in.) = 3.78 in.^{2}

A–C–E: 4 in.^{2} – (3)(.75 in.)(.5 in.) + (2)(.90)(.5 in.) = 3.78 in.^{2}

The critical (smallest) net area is across section A–B and is 3.25 in.^{2}.

The allowable load across the effective net area (using net area = 3.25 in.^{2} and F_t = 29 ksi) can be calculated as follows:

P_{all} = 3.25 in.^{2} × 29 ksi = 94.3 kips

Since 94.3 kips is greater than the actual load of 50 kips, the member is adequate across the effective net area.

Since the member meets AISC criteria across both gross and effective net area, the member is adequate.

4.6   Pin-Connected Members and Rods

Many older truss bridges in use today have pin-connected members, instead of bolted or welded gusset plates. This configuration was commonly used on bridges until the advent of bolting and welding. Today, pin-connected members may still be used for special structures such as utility hangers, but, in reality, they are not used very often. Pin-connected tension members have different allowable stress considerations, as found in Section D3 of the AISC manual. These cases will be briefly outlined in the following paragraphs.

The allowable stress on the net area of the pin hole of a tension member is given as 0.45 F_y, although bearing stresses on the projected area must also conform to AISC criteria. These bearing stresses will be further discussed in Chapter 9.

Threaded rods are also used in many hanger-type applications, from suspended walkways to sag rods in roof systems. The allowable stress that the AISC specifies over the major diameter of a rod is $0.33\ F_u$. This allowable stress ($0.33\ F_u$) is the criterion to prevent fracture, but the student should also check to make sure that the capacity against yielding along the body of the rod does not control. The allowable stress against gross area yielding along the rod is again listed as $0.60\ F_y$.

The following example will illustrate the use of the AISC specifications as they pertain to threaded rods.

EXAMPLE 4.6

Design the diameter of a threaded rod to hold a load of 15 kips. The steel is A36 (F_y = 36 ksi, and F_u = 58 ksi).

(Designing the diameter of a rod is actually a function of area. Therefore, design per the fracture criteria, and then check this design against the yielding criteria over the gross area of rod.) Design area based on fracture:

$A_{min} = P/F_t$ (Since load is given at 15 kips and the F_t = .33 (58 ksi) = 19.1 ksi)

A_{min} = 15 kips/19.1 ksi = .785 in.^{2}

Since the rod is circular, $A = \pi/4 \times (d)^2$ and solving for diameter gives us a diameter equal to 1 in.

Check a 1-in.-diameter rod for gross area capacity:

A = .785 in.^{2}

P = 15 kips

Therefore, the actual stress is

f_t = 15 kips/.785 in.^{2} = 19.1 ksi < $0.60\ F_y$.

Since the 1-in. rod meets both criteria, it works.

4.7   Summary

Tension members are relatively common in steel construction, being utilized in trusses, frames and various cable and rod applications. The ASD method specifies that the tension member's allowable stress (F_t) is greater than or equal to the actual tensile stress (f_t).

The tension member being considered must always be checked or designed over the gross area and the effective net area. The gross area is the section's full cross-sectional area, while the effective net area is the section with the bolt holes taken out. Sometimes the effective net area is reduced further to account for the ineffectiveness in transferring stress to adjoining members. The failure mode which is probable over the gross area is yielding while that over the effective net area is fracture.

EXERCISES

1. Find the maximum allowable load permitted on a 6-in.-wide × 3/4-in.-thick plate with 1/2-in.-diameter bolts, as shown in the figure. The steel is A36.

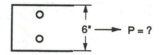

2. Find the stress in the gross and net area for an 8-in.-wide × 3/4-in.-thick plate subjected to an axial load of 55 kips. The plate has three 3/4-in.-diameter bolt holes, and the steel is A36. Are these stresses acceptable per AISC criteria?

3. Design the thickness of an 8-in.-wide steel plate with the bolt layout as shown in the figure. The bolts are 3/4 in. in diameter, and the steel is A242 ($F_y = 50$ ksi, and $F_u = 70$ ksi).

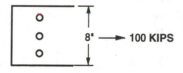

4. The W 12 × 50 shown in the figure is used as a tension member and is bolted through its flanges, as shown, with 3/4-in.-diameter bolts. Determine the maximum allowable load. The steel is A36.

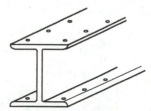

5. Find the tension stress over the gross and effective net areas for a 4 × 4 × 1/2 angle connected as shown by the 5/8-in.-diameter bolts. The angle has a load of 45 kips, and the steel is A36. Compare stresses to AISC allowable stresses. Does this angle work?

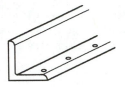

6. Determine the critical net area through the plate shown in the figure. The plate is 8 in. wide × 1 in. thick, and the bolts are 1/2 in. in diameter.

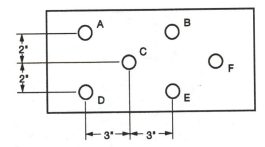

7. If the plate in Exercise 6 has a 160-kip tensile force on it, is it adequate per AISC standards?
8. Design the most economical W12 section to hold a tensile load of 400 kips. The bolts and connection details are the same as in Exercise 4. The steel is A36.
9. Design the most economical 4 × 4 angle to hold a tensile load of 135 kips. The angle has one line of 3/4-in.-diameter A490 bolts with 3 bolts in this line. The steel is A242 (F_y = 50 ksi, and F_u = 70 ksi).
10. Calculate the allowable load that can be placed on a 1 1/8-in.-diameter threaded rod, if the rod is made from A36 steel. What is this rod's capacity over the gross area of its body?
11. Calculate the allowable load on the 7 × 4 × 1/2 double angles placed long legs back to back. The bolts are 3/4 in. diameter and are placed as shown in the figure. The steel is A36.

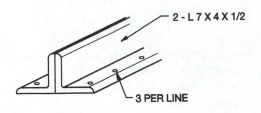

REFERENCES

1. Gerner A. Olsen, *Elements of Mechanics of Materials* (Englewood Cliffs, N.J.: Prentice-Hall, 1974), p. 520.
2. Patrick Dowling, Peter Knowles, and Graham Owens, *Structural Steel Design* (London: The Steel Construction Institute, 1988), p. 156.
3. Charles G. Salmon and John E. Johnson, *Steel Structures Design and Behavior* (New York, Harper and Row, 1980), p. 70.

5

CONCENTRIC COMPRESSION MEMBER DESIGN

5.1   Introduction

Compression members are found in all types of construction, from the skeleton framing of a building to the massive pier towers of the Golden Gate Bridge (Figure 5–1). If compression members are set vertically, they are commonly referred to as **columns**. The student should note that the terms *compression members* and *columns* will be used interchangeably throughout this chapter.

The importance of compression members cannot be underestimated. They are extremely important in the overall design in steel buildings or bridges. The student must remember that if a column fails, everything supported above that column will most probably collapse.

This chapter will present the AISC technique for concentric compression member design using the Allowable Stress Design method. Discussion will include column instability and the factors relating this behavior to the AISC equations.

It should be stated that columns are rarely, if ever, loaded concentrically in the real world. More typically, they are subjected to some bending moment either through eccentric axial loads, beam reactions, lateral loadings, or a combination thereof. When such situations induce bending stresses into the axially loaded compression member, these members are typically referred to as **beam-columns**. This topic will be the focus of Chapter 8.

Figure 5–1   The Golden Gate Bridge, San Francisco. (Courtesy Bethlehem Steel Corporation.)

5.2   Potential Modes of Column Failure

Column behavior is notably different from that of tension members. Whereas the tension member under stress tends to "straighten out" in the direction of the load, a column tends to "move out" of the plane of loading. This tendency to "move out" is referred to as **buckling** and is a serious concern in column design. Buckling constitutes failure because the column is unstable and cannot accept any additional load.

Another method of column failure is the yielding (crushing) of the material due to the actual stress on the member being greater than the yield. It is evident that as columns get longer, the dominating failure mode is buckling. Only very short columns will fail due to yielding, and the intermediate length column will fail by a combination of these two types of behavior (Figure 5–2).

This intermediate behavior occurs because these columns do not buckle until stresses have reached a sufficiently high level, thereby initiating some yield-

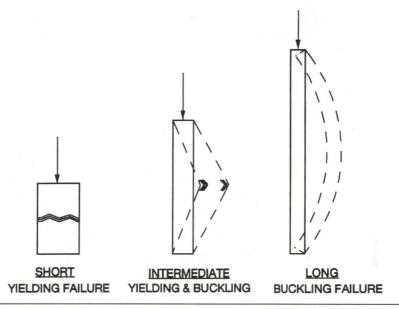

SHORT
YIELDING FAILURE

INTERMEDIATE
YIELDING & BUCKLING

LONG
BUCKLING FAILURE

Figure 5–2   Various Failure Modes Relative to Column Length.

ing of the fibers. The vast majority of columns fall in the category of intermediate behavior.

5.3   Column Behavior

The understanding of columns and the use of today's design equations can be traced back to a Swiss mathematician, Leonhard Euler. In the mid-eighteenth century, Euler studied column buckling and derived a formula to predict the load that would cause a column with rounded ends to buckle.[1] This formula is the basis of all modern-day column equations, and is as follows:

$$P = \frac{\pi^2 EI}{l^2}$$

where

P = buckling load

E = material's modulus of elasticity

I = moment of inertia

l = length of column between points of zero moment

Usually this formula is written in terms of buckling stress (P/A) by substituting Ar^2 for the moment of inertia, I. We can then express Euler's equation in terms of buckling stress as follows:

$$\frac{P}{A} = \frac{\pi^2 E}{(l/r)^2}$$

where r = radius of gyration about the buckling axis.

From Euler's equation the student can see that buckling is dependent on basically two factors: the l/r term (typically referred to as slenderness ratio) and the column's modulus of elasticity. The student should study Euler's equation and realize that as the slenderness ratio (l/r) increases, the buckling stress decreases. That is, as a column gets longer (therefore, more "slender"), the stress to initiate buckling becomes smaller.

Because of the incorporation of the modulus of elasticity in the Euler equation, its use was limited by the fact that it would describe only buckling that occurred in the material's elastic region. When buckling in the elastic region occurs, it is referred to as **elastic buckling**. This phenomenon of elastic buckling is limited to long, slender columns that buckle under low stress levels.

This limitation made the equation accurate in predicting buckling behavior for only a small number of columns. Columns of intermediate length do not buckle elastically; therefore, Euler's equation tended to overestimate their capacity (Figure 5–3). Buckling for intermediate-length columns occurs at higher stress levels, a

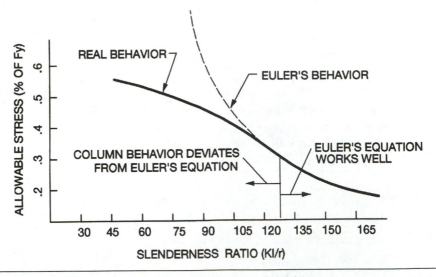

Figure 5–3   Actual Column Behavior Versus Euler Behavior.

type of buckling referred to as **inelastic buckling,** since some fibers will have yielded as buckling commences. Because many typical columns are of intermediate length, the Euler equation was ignored for many years[2] while engineers searched for a single equation to describe buckling over all column lengths.

A single column equation defining column behavior over all ranges of slenderness ratios has never been found. In lieu of this absence, modern-day column design equations are based on the fact that two types of buckling occur—elastic and inelastic. Elastic buckling occurs as stresses on a column are rather low but still under the material's proportional limit. Inelastic buckling occurs at higher stress levels and cannot be well defined by the original Euler equation. The AISC recognizes this fact and proposes therefore that two column equations (one for elastic and one for inelastic behavior) be used in design. We will discuss these AISC equations further in Section 5.5.

5.4 Effective Length

The amount of fixity existing at the ends of a column is known to affect a column's buckling resistance dramatically. Thus, there has been some modification of Euler's original equation, since his testing dealt only with one type of end-restraint condition. Actually, columns may have many types of end-restraint conditions, so a modification factor that estimates effective length of the column has been introduced. This modification factor is referred to as the **effective length factor,** K.

This effective length factor approximates the length over which a column actually buckles. This buckling length can be longer or shorter than the actual length of the column, based on end restraints of the column and/or whether the column is subjected to lateral movement. The student should realize that as the length of a column is reduced, the column actually increases its resistance to buckling; conversely, as the column length is increased, the resistance to buckling decreases. End-restraint conditions on a column will dramatically influence the buckling length, or **effective length,** Kl, of this member. The effective length factor, K, is an attempt to describe accurately the true buckling length of the column.

Much research has been devoted to the investigation of K factors in both braced and unbraced frames. The chart in Figure 5–4 contains widely recognized K values for *ideal* end-restraint conditions (pin, fixed, free, and so on). The AISC recognizes that in real practice, there are no ideal conditions; therefore, recommended K values are also shown. Notice that the recommended values are always greater than or equal to the theoretical values.

The following example illustrates the power of the effective length factors as they are taken from this chart. Keep in mind that a column's effective length and its allowable stress have an inversely proportional relationship. That is, as a column becomes "shorter," the calculated allowable stress will increase.

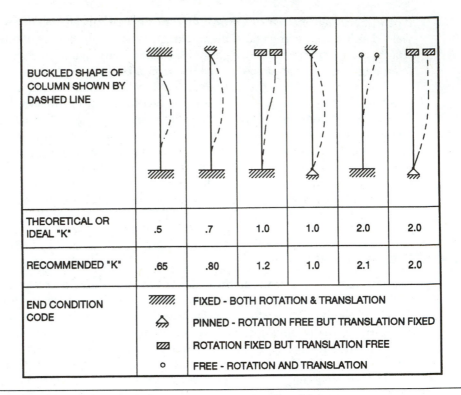

BUCKLED SHAPE OF COLUMN SHOWN BY DASHED LINE						
THEORETICAL OR IDEAL "K"	.5	.7	1.0	1.0	2.0	2.0
RECOMMENDED "K"	.65	.80	1.2	1.0	2.1	2.0
END CONDITION CODE		FIXED - BOTH ROTATION & TRANSLATION				
		PINNED - ROTATION FREE BUT TRANSLATION FIXED				
		ROTATION FIXED BUT TRANSLATION FREE				
	o	FREE - ROTATION AND TRANSLATION				

Figure 5–4   Typical Values of *K* for Idealized End Conditions.

EXAMPLE 5.1

Calculate the effective length, *Kl*, of a 20-ft-long column under the following three end-restraint conditions. Use the recommended values from the chart.
   1.   pin–pin
   2.   pin–fixed
   3.   fixed–fixed

The effective length should be calculated as follows.

1.   Since the recommended *K* value for a pin–pin end condition is 1.0, the effective length, *Kl*, of this value is 1.0 × 20 ft = *20 ft.*

2.   The recommended *K* value for a pin–fixed end condition is .80; therefore, the effective length is .80 × 20 ft = *16 ft.* (Essentially because of its end conditions, this column behaves similarly to a 16-ft-long column with pin–pin end conditions.)

3.   With the recommended *K* value equal to *.65*, the effective length is equal to *.65 × 20 ft = 13 ft*. (Again, the column has essentially the same behavior as a 13-ft-long column with pinned ends.)

5.5   AISC Column Design Philosophy Using ASD

From the time of Euler's original buckling equation, there have been only two modifications that every design buckling equation now incorporates. The first, which was mentioned in Section 5.4 is the use of the effective length factor, *K*, to estimate more accurately the true buckling length of a column. The second modification is the use of a safety factor, since the allowable stress philosophy must, in all cases, ensure a proper margin of safety.

The AISC formulas for column design calculate the allowable axial stress permitted on a column and are the result of years of research committed to finding a formula that accurately predicts behavior in all slenderness ranges. Many formulas have been proposed, all realizing that the slenderness ratio of the column dictates whether elastic or inelastic buckling will occur. The AISC uses a parabolic formula in the inelastic range and a modified Euler formula for elastic behavior.

The AISC philosophy recognizes that the behavior of a column changes from the elastic to the inelastic range. The AISC therefore sets a value of a slenderness parameter, called C_c, as the dividing line between the two aforementioned ranges. The value of this slenderness parameter (C_c) is actually the slenderness ratio that would cause the buckling stress to be 50% of a steel's yield stress by using Euler's equation. The reason for using a value of only 50% of F_y to delineate elastic from inelastic buckling is due to the effect of residual stress in rolled members. Residual stresses are those stresses that are "built in" rolled members due to uneven cooling during the manufacturing process.

Residual stresses can be as high as 20%–30% of F_y and exist in steel members before they are ever loaded! These stresses can lead to the buckling of intermediate columns at stresses below their theoretical critical load;[3] therefore, the AISC specification uses a level of 50% of F_y as the boundary between elastic and inelastic behavior.

The value of C_c can be calculated as follows:

$$C_c = \sqrt{\frac{2\pi^2 E}{F_y}} \qquad \text{(AISC specification, page 5–42)}$$

The AISC uses C_c to distinguish whether elastic or inelastic behavior will occur in a steel column. When a column's individual slenderness ratio, *Kl/r*, is less than C_c, inelastic buckling predominates. In a similar manner, when a column's slenderness ratio, *Kl/r*, is greater than C_c, elastic buckling controls.

When a column is controlled by inelastic buckling, the AISC uses the parabolic formula (AISC Eq. E2–1, page 5–42) to predict its allowable stress accurately . This formula is as follows:

$$F_a = \frac{\left[1 - \dfrac{(Kl/r)^2}{2C_c^2}\right]F_y}{\dfrac{5}{3} + \dfrac{3(Kl/r)}{8C_c} - \dfrac{(Kl/r)^3}{8C_c^3}} \qquad \text{(AISC Eq. E2–1)}$$

Although this equation looks rather imposing, the denominator is actually just its safety factor, which can theoretically range from 5/3 (when the column's $Kl/r = 0$) to 23/12 (when the column's $Kl/r = C_c$). This variable safety factor increases as the column becomes more slender to account for the detrimental effects caused by possible column crookedness.

For a column controlled by elastic behavior, the AISC recognizes the validity of Euler's equation and therefore uses the same formula, modified only by a safety factor and the effective length factor, K. The safety factor for columns in the elastic range is given as 23/12 (≈ 1.92). When the slenderness ratio, Kl/r, of an individual

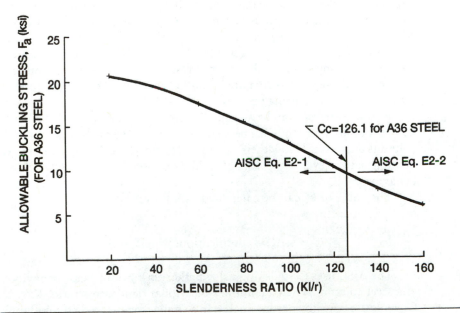

Figure 5–5   Use of AISC Column Equations Relative to Behavior.

column exceeds C_c, the AISC uses the following equation (AISC Eq. E2–2, page 5–42) to describe the allowable buckling stress:

$$F_a = \frac{12\pi^2 E}{23(Kl/r)^2} \qquad \text{(AISC Eq. E2–2)}$$

The student should note the striking similarity between this equation for allowable stress and the 250-year-old Euler equation for actual buckling stress.

The AISC philosophy is easily seen in Figure 5–5. The following example will illustrate the calculation of allowable stress per AISC formulas.

EXAMPLE 5.2

Calculate the allowable stress for columns having slenderness ratios of 60, 85, and 140. The columns are all made from A36 steel (F_y = 36 ksi). The modulus of elasticity is 29,000 ksi.

First calculate the value of C_c, since this will set the dividing line between the inelastic and elastic ranges (i.e., whether we use AISC Eq. E2–1 or E2–2 in the calculation of allowable stress).

$$C_c = \sqrt{2\pi^2(29000)/36} = 126.1$$

For a Kl/r = 60, since it is less than C_c, use AISC Eq. (E2–1) (using 60 and 126.1 for Kl/r and C_c, respectively):

$$F_a = 17.43 \text{ ksi}$$

For a Kl/r = 85, since it is also less than C_c, use AISC Eq. (E2–1) (using 85 and 126.1 for Kl/r and C_c, respectively):

$$F_a = 14.79 \text{ ksi}$$

For a Kl/r = 140, since this is greater than C_c, use AISC Eq. (E2–2) (using 140 for Kl/r and 29,000 ksi for E):

$$F_a = 7.62 \text{ ksi}$$

(Note how the allowable stress decreases as the column becomes more slender.)

Once students have mastered the rather tedious calculations involved in solving AISC Eq. (E2–1), they will be happy to find that there is a very helpful aid to short-cut this procedure. The AISC provides Table 3 on page 5–119 of its specification, which is reproduced here in Table 5–1. This table has precalculated ordinates (referred to as C_a) along the curve for allowable stress in the inelastic region (when $Kl/r < C_c$). The student simply enters the table using the ratio of slenderness ratio to C_c ($Kl/r/C_c$) and selects the appropriate value of C_a. The value of C_a is then multiplied by the steel's yield stress (F_y) to attain the correct value of allowable stress. Students are urged to rework the problem in Example 5.2 using this method, to familiarize themselves with this design aid.

Table 5–1   Values of C_a: For Determining Allowable Stress When $Kl/r \le C_c$ for Steel of Any Yield Stress (by Eq. $F_a = C_a F_y$)*

$\dfrac{Kl/r}{C_c}$	C_a	$\dfrac{Kl/r}{C_c}$	C_a	$\dfrac{Kl/r}{C_c}$	C_a	$\dfrac{Kl/r}{C_c}$	C_a
.01	.599	.26	.548	.51	.472	.76	.375
.02	.597	.27	.546	.52	.469	.77	.371
.03	.596	.28	.543	.53	.465	.78	.366
.04	.594	.29	.540	.54	.462	.79	.362
.05	.593	.30	.538	.55	.458	.80	.357
.06	.591	.31	.535	.56	.455	.81	.353
.07	.589	.32	.532	.57	.451	.82	.348
.08	.588	.33	.529	.58	.447	.83	.344
.09	.586	.34	.527	.59	.444	.84	.339
.10	.584	.35	.524	.60	.440	.85	.335
.11	.582	.36	.521	.61	.436	.86	.330
.12	.580	.37	.518	.62	.432	.87	.325
.13	.578	.38	.515	.63	.428	.88	.321
.14	.576	.39	.512	.64	.424	.89	.316
.15	.574	.40	.509	.65	.420	.90	.311
.16	.572	.41	.506	.66	.416	.91	.306
.17	.570	.42	.502	.67	.412	.92	.301
.18	.568	.43	.499	.68	.408	.93	.296
.19	.565	.44	.496	.69	.404	.94	.291
.20	.563	.45	.493	.70	.400	.95	.286
.21	.561	.46	.489	.71	.396	.96	.281
.22	.558	.47	.486	.72	.392	.97	.276
.23	.556	.48	.483	.73	.388	.98	.271
.24	.553	.49	.479	.74	.384	.99	.266
.25	.551	.50	.476	.75	.379	1.00	.261

Source: Courtesy of the American Institute of Steel Construction, Inc.

*When ratios exceed the noncompact section limits of AISC Sect. B5.1, use $\dfrac{Kl/r}{C'_c}$ in lieu of $\dfrac{Kl/r}{C_c}$ values and equation $F_a = C_a Q_a Q_s F_y$ (Appendix Sect. B5).

Note: Values are for all grades of steel.

5.6   Braced Columns

We have already discussed how columns will behave as the stress on them becomes increasingly large. So far in design, we realize that to carry more load, a member's area can always be increased. But can we increase the load-carrying capacity of these columns without using a larger rolled section?

Yes, with columns we can accomplish an increase in strength by reducing the effective length, Kl, through the use of bracing. Bracing of columns can take various forms, such as steel X-bracing, the framing in of beams, or the bracing effects of a floor system. All of these techniques will reduce the overall buckling length of column, and as this length is reduced, the stress that the column is allowed to carry will increase (Figure 5–6).

The location of this bracing is also very important to the load-carrying capacity of the column. The student will remember that the buckling load formulas all incorporate the Kl/r term known as the slenderness ratio. If we look again at the AISC buckling equations, we see that the only variable (once a grade of steel and length of column have been decided on) is the radius of gyration, r. The radius of gyration is an elastic property of area for an individual rolled section based on the axis under consideration and is given by the formula:

$$r = \sqrt{I/A}$$

where

I = moment of inertia about the axis being considered

A = the cross-sectional area of the shape

As students view the section property tables, they will notice that r_y is always less than r_x for wide-flange sections. Therefore, for an unbraced wide-flange column, the

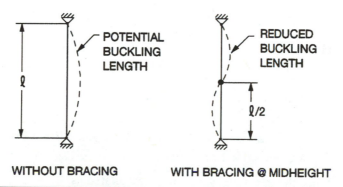

Figure 5–6   Bracing Effect on Column Buckling.

slenderness ratio Kl/r_y will always be larger than Kl/r_x (assuming that K remains the same about both axes). This means that buckling of an unbraced column will always occur about the weak or y–y axis because the column is more slender in this direction. By bracing the column in the weak direction (y–y), we also reduce the Kl/r_y value to produce a larger allowable stress. By lowering Kl/r_y, we might even reach the point where the weak axis no longer controls ($Kl/r_y < Kl/r_x$).

The design equations for columns with bracing are exactly the same as outlined in the previous section, although when bracing is used, the student cannot assume that buckling will necessarily occur about the weak axis. **When bracing is used or when the effective length factor is different between the x and y axes, the slenderness ratio about both the strong and the weak axes must be calculated. The higher of these two slenderness ratios will control the design, since the higher of the two will lead to a smaller allowable stress.**

5.7 Column Design Problems per ASD

As discussed in previous chapters, two general types of problems are encountered in the design arena: evaluation and design. In the evaluation problem, you will be given a member and corresponding load and asked to determine its adequacy or to determine how much load it can safely withstand per the specification. For the design problem, you will be given loads and asked to design the member size based on your knowledge of the applicable specification criteria using allowable stress techniques.

The evaluation problem is very easy to perform; the following examples will demonstrate how the typical evaluation problem is performed. Following these evaluation examples, a technique for solving the column design problem will be presented.

EXAMPLE 5.3

Calculate the allowable load, P_{all}, which an unbraced W 14 × 53 column can hold if it is 20 ft long with $K = 1.0$ (in both directions). The steel is A36.

A W 14 × 53 has $A = 15.6$ in.^{2}.

$r_x = 5.89$ in.

$r_y = 1.92$ in.—r_y will control since column is not braced.

$Kl_y/r_y = 1.0 \times 20$ ft $\times 12$ in./ft/1.92 in. $= 125$.

Since the steel is A36, C_c can be calculated as

$$\sqrt{2\pi^2(29,000)/36} = 126.1$$

Since our controlling $Kl/r < 126.1$, use AISC Eq. (E2–1) to calculate the allowable stress, F_a. Calculating F_a yields 9.55 ksi. Therefore, the allowable load, P_{all}, is

$$P_{all} = 9.55 \text{ ksi} \times 15.6 \text{ in.}^2 = 149 \text{ kips}$$

(Remember, when bracing is used, the slenderness ratios about both the strong and weak axes should be checked, with the larger ratio controlling the design.)

EXAMPLE 5.4

(a) Calculate the allowable load, P_{all}, for a W 24 × 104 that is 20 ft long with $K = .80$ (for both axes). (b) Then, recalculate if the column is braced at midheight in the weak axis (still assume $K = .80$). Use A36 steel.

a.   W 24 × 104 $A = 30.6 \text{ in.}^2$

$r_x = 10.1$ in.

$r_y = 2.91$ in.

Kl_y/r_y controls = .8(20 ft × 12 in./ft)/2.91 in. = 65.98.

Since the column is made from A36 steel, $C_c = 126.1$. Therefore, because the controlling $Kl/r_y < C_c$, use AISC Eq. (E2–1) to calculate the allowable stress, F_a. $F_a = 16.84$ ksi and the allowable load is as follows:

$$P_{all} = 16.84 \text{ ksi} \times 30.6 \text{ in.}^2 = 515.3 \text{ kips}$$

b.   Braced midheight in weak direction:

Kl_x/r_x = .8(20 ft × 12 in./ft)/10.1 in. = 19.01

Kl_y/r_y = .8(10 ft × 12 in./ft)/2.91 in. = 32.99 still controls since it is the larger slenderness ratio. Again, from AISC Eq. (E2–1), calculate the allowable stress, F_a, as equal to 19.73 ksi. The allowable load is then calculated as follows:

$$P_{all} = 19.73 \text{ ksi} \times 30.6 \text{ in.}^2 = 603.7 \text{ kips}$$

(The student should notice that bracing increases capacity by (603.7 – 515.3)/ 515.3 = 17.1%.)

EXAMPLE 5.5

An unbraced W 12 × 79 has 280 kips of axial compressive load applied to it. It is 20 ft long, made from A36 steel and has pin–pin end conditions (same for both axes). Is it okay per AISC criteria?

In this problem, we will evaluate this column by calculating the actual axial stress and comparing that value to the allowable axial stress.

The actual axial stress is

$$f_a = 280 \text{ kips}/23.2 \text{ in.}^2 = 12.07 \text{ ksi}$$

The allowable stress is based on the controlling slenderness ratio, which is Kl_y/r_y because the column is unbraced. This value is 1.0(20 ft × 12 in./ft)/ 3.05 in. = 78.69 (Where $K = 1.0$, $l = 20$ ft, and $r_y = 3.05$ in. for this W 12 × 79.) Since 78.69 < 126.1 (which is C_c), use AISC Eq. (E2–1) (or enter Table 5–1 with $Kl_y/r_y/C_c$ = to 78.69/126.1 = .624) to calculate the allowable axial stress, F_a. Calculating, we find that $F_a = 15.49$ ksi. Since our actual stress on this column ($f_a = 12.07$ ksi) is less than 15.49 ksi, the column will be adequate.

Now that we have had sufficient exposure to the evaluation-type problem, we are ready to focus on true column design. In design, the author believes it is best to proceed as though students will not have the many existing design tables at their disposal. (Later in this section we will explore the utilization of the AISC column design tables that are commonly used.) The "from-scratch" design procedure that the author presents is actually a trial-and-error approach, but it is found to be very quick if students utilize their knowledge of column behavior. This procedure also helps to underscore the importance of learning the behavior of columns to assist in the overall understanding of design.

The realization that the allowable stress, F_a, is dependent only on the slenderness ratio of a column is the key point that students must grasp. As they become more experienced in column design and behavior, students will be able to estimate values of allowable stress based only on their familiarity with a general range of slenderness ratio. A step-by-step procedure for column design may be outlined as follows:

1. Estimate a value of allowable stress, $F_{a \text{ est}}$, based on the controlling slenderness ratio. (The accuracy of this step depends greatly on the designer's experience.)
2. Using $F_{a \text{ est}}$, calculate the required area as $A_{\text{req'd}} = P/F_{a \text{ est}}$.

3.   Choose a trial member based on $A_{req'd}$. Realize that $A_{req'd}$ is only a "ballpark" area and is only as good as your estimate of allowable stress.
4.   Calculate the allowable stress, F_a, of the trial member based on its controlling slenderness ratio. Use AISC Eq. (E2–1) or (E2–2) (use Table 5–1 when $Kl/r < C_c$; it is much faster).
5.   Calculate the allowable load, P_{all}, of the trial member as follows: $P_{all} = A_{\text{trial member}} \times F_a$.
6.   If the allowable load of the trial member is equal to or just a little greater than the actual load, the member is good. If the allowable load of the trial member is less than the actual load, the member is no good. If the allowable load of the trial member is much greater than the actual load, the member works, but there is probably a smaller and more economical member available.

The following examples will introduce this design procedure. Remember to follow logic when sizing up or down after the first trial is made.

EXAMPLE 5.6

Find the most economical W12 section to hold a compressive load of 120 kips. The steel is A36, $K = 1.0$, and $l = 16$ ft (for both axes). The column is unbraced.

Since the column is unbraced, we know that the weak axis slenderness ratio controls the design. With an effective length, Kl, equal to $1.0 \times (16 \text{ ft} \times 12 \text{ in./ft})$ = 192 in., let's take a look at the r_y values for W12 sections. Notice that, except for the very small sections, most r_y values range from ≈ 2 to ≈ 3.5. If we assume an average r_y value of 2.75, that would make our slenderness ratio ≈ 192 in./2.75 in. ≈ 69.8. This value shows that this column will most probably be controlled by inelastic buckling since $69.8 < C_c$ (126.1 for A36 steel). With this assumed slenderness ratio in the inelastic region, we can start the design process by estimating a rather high value of allowable stress.

1.   Estimate $F_a = 15$ ksi.
2. $A_{req'd} = 120$ kips/15 ksi = 8 in.^{2}.
3.   From the tables, choose a trial member of W 12×30 ($A = 8.79$ in.^{2}).
4.   Calculate the allowable stress, F_a, for the trial section of W 12×30. The member has a Kl/r_y of 1.0(16 ft $\times$ 12 in./ft)/1.52 in. = 126.3, which is greater than C_c, so use AISC Eq. (E2–2). Find $F_a = 9.36$ ksi.
5.   The allowable load, P_{all}, for this W 12×30 is simply = 9.36 ksi $\times$ 8.79 in.^{2} = 82.3 kips.

6. Since allowable load of 82.3 kips < 120 kips, this section does not work. Our original estimate of F_a was probably too high. Start again with $F_{a\,est} =$ 10 ksi.

1. $F_{a\,est} = 10$ ksi.
2. $A_{req'd} = 120$ kips/10 ksi = 12 in.^{2}.
3. Choose W 12 × 45 ($A = 13.2$ in.^{2}) as the new trial member.
4. Calculate the allowable stress, F_a, for this W 12 × 45. The controlling slenderness ratio, Kl/r_y, equals 192 in./1.94 = 99.0, which is less than C_c. Therefore, by using AISC Eq. (E2–1) or Table 5–1, we find $F_a = 13.1$ ksi.
5. The allowable load for this trial member is 13.1 ksi × 13.2 in.^{2} = 173 kips.
6. Since the allowable load (173 kips) is greater than the actual load (120 kips), the member works. Since the allowable load is somewhat larger than the actual, it would be wise to check a section which was a little smaller. (A W 12 × 40 is the best.)

(If there is ever a doubt that a section is actually the smallest that works, the student should calculate the allowable load for the next smallest section in the table.)

In addition to the "from-scratch" procedure outlined previously, the AISC provides column load tables in Part 3 of its ASD manual. (A partial listing of these load tables is found in Appendix C.) These tables contain much information, the foremost of which is the allowable concentric load that can be carried based on a given effective length (Kl) about the weak axis on a given wide-flange section. Example 5.7 is a repeat of Example 5.6, this time utilizing the column load tables.

EXAMPLE 5.7

Using the column load tables, find the most economical W12 section to hold a compressive load of 120 kips. Steel is A36, $K = 1.0$, $l = 16$ feet. The column is unbraced.

Since the column is unbraced, we can calculate the effective length about the weak axis only, since this will be the axis of buckling.

$Kl = 1.0(16\ \text{ft}) = 16\ \text{ft}$

Entering the column load tables for W12 sections having a $Kl = 16$ ft about the weak axis, we notice that a W 12 × 40 is the smallest section available in the tables and that this section has an allowable load of 154 kips at this length. (The "light" column is used in this case because the steel has an $F_y = 36$ ksi; should the steel have an $F_y = 50$ ksi, the "dark" column would be used.)

F_y = 36 ksi									
F_y = 50 ksi									

COLUMNS
W shapes
Allowable axial loads in kips

Designation	W12									
Wt./ft	58		53		50		45		40	
F_y	36	50	36	50	36	50	36	50	36	50‡
0	367	510	337	468	318	441	285	396	255	354
6	341	464	312	425	286	386	256	346	229	309
7	335	454	307	416	279	374	250	335	223	299
8	329	443	301	406	271	360	243	322	217	288
9	322	432	295	395	263	346	235	309	210	276
10	315	420	288	384	254	331	228	296	203	264
11	308	407	282	372	246	315	220	281	196	251
12	301	394	275	360	236	298	211	266	188	237
13	293	380	268	347	226	281	202	250	180	222
14	285	365	260	333	216	262	193	233	172	207
15	276	351	252	319	206	243	183	216	163	191
16	268	335	244	305	195	223	173	197	154	175
18	249	302	227	274	171	181	152	159	135	141
20	230	267	209	241	146	146	129	129	114	114
22	209	229	189	206	121	121	106	106	94	94
24	187	193	169	173	102	102	89	89	79	79
26	164	164	147	147	87	87	76	76	67	87
28	142	142	127	127	75	75	66	66	58	58
30	123	123	111	111	65	65	57	57	51	51
32	108	108	97	97	57	57	50	50	45	45
34	96	96	86	86						
38	77	77	69	69						
41	66	66	59	59						

Leftmost column label: Effective length in ft KL with respect to least radius of gyration r_y

Properties										
U	3.21	3.21	3.24	2.94	4.10	4.10	4.12	3.75	3.77	3.77
P_{wo} (kips)	89	124	78	108	92	127	75	105	66	92
P_{wi} (kips/in.)	13	18	12	17	13	19	12	17	11	15
P_{wb} (kips)	121	142	106	125	131	155	97	115	66	78
P_{fb} (kips)	92	128	74	103	92	128	74	103	60	83
L_c (ft)	10.6	9.0	10.6	9.0	8.5	7.2	8.5	7.2	8.4	7.2
L_u (ft)	24.4	17.5	22.0	15.9	19.6	14.1	17.7	12.8	16.0	11.5
A (in.^{2})	17.0		15.6		14.7		13.2		11.8	
I_x (in.^{4})	475		425		394		350		310	
I_y (in.^{4})	107		95.8		56.3		50.0		44.1	
r_y (in.)	2.51		2.48		1.96		1.94		1.93	
Ratio r_x/r_y	2.10		2.11		2.64		2.65		2.66	
B_x } Bending	0.218		0.221		0.227		0.227		0.227	
B_y } factors	0.794		0.813		1.058		1.065		1.073	
$a_x/10^6$	70.6		63.6		58.8		52.2		46.3	
$a_y/10^6$	16.0		14.3		8.4		7.4		6.5	
$F'_{ex} (K_xL_x)^2/10^2$ (kips)	289		284		278		275		273	
$F'_{ey} (K_yL_y)^2/10^2$ (kips)	65.3		63.8		39.8		39.0		38.6	

‡Web may be noncompact for combined axial and bending stress;
see AISC ASD Specification Sect. B5.1.
Note: Heavy line indicates Kl/r of 200.

Since the allowable load of 154 kips is greater than the actual load of 120 kips, the section is definitely okay. (A W 12 × 35 could be checked to see whether it might also be okay, but we would find that it does not work.)

5.8   Local Buckling

Another concern with regard to compression members is that the individual components of the section do not fail before the member reaches its capacity. Buckling of individual pieces of the compression member (the web and/or flanges) before the whole section buckles is referred to as local buckling. In Section B.5 of the AISC specification, the criteria to prevent local buckling for stiffened and unstiffened compression elements are set. An element can meet the AISC criteria against local buckling in one of two ways: either by being **compact** or **noncompact**. An element is said to be compact if the section can reach significant plastic deformation before buckling would occur, and it is said to be noncompact if the section can reach at least its yield strength before buckling would occur. We will further discuss this topic of compactness in Chapter 6.

The AISC specification categorizes stiffened elements as those supported on two edges parallel to the load, and unstiffened elements as those supported only along one edge. For a standard wide-flange shape, the flange is assumed to be an unstiffened element, since one-half of it "cantilevers" out from its point of fixity at the web. The web is assumed to be a stiffened element, since it is fixed at its junction with both flanges (Figure 5–7).

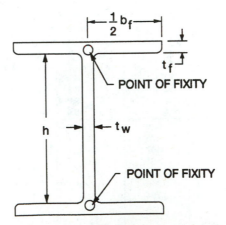

STIFFENED ELEMENT = WEB; WIDTH/THICKNESS RATIO = h/t_w
UNSTIFFENED ELEMENT = FLANGE; WIDTH/THICKNESS RATIO = $1/2\ b_f/t_f$

Figure 5–7   Stiffened and Unstiffened Elements on a Rolled Beam.

Table 5–2   Compactness Limits for Wide-Flange Sections in Compression

		Limiting b/t Ratios	
Description of Element	b/t Ratios	Compact	Noncompact
Flanges of I-shaped rolled beams and channels in flexure	$1/2\ b_f/t_f$	$65/\sqrt{F_y}$	$141/\sqrt{F_y}-10$
Flanges of I-shaped sections in pure compression and plates projecting from compression elements	$1/2\ b_f/t_f$	NA	$95/\sqrt{F_y}$
All other uniformly compressed stiffened elements	h/t_w	NA	$253/\sqrt{F_y}$

NA = Not applicable.

The stiffened and unstiffened portions of a wide-flange section behave similarly to a column because as their "slenderness ratio" becomes smaller it is harder for the element to buckle. For local buckling we refer to this "slenderness ratio" as the element's width-to-thickness ratio, or b/t ratio. The AISC has set limits for an element's b/t ratio to ensure that local buckling will not occur before the section fails due to compact or noncompact behavior. These limits are outlined in Table 5–2.

All rolled wide-flange sections subject to axial compression meet the compactness criteria shown in Table 5–2, and local buckling on these elements is therefore practically nonexistent. However, when dealing with plate girders or other builtup sections, the designer must be aware that the allowable stress may have to be reduced to ensure safety against this mode of failure.

5.9   Alignment Charts for Determining Effective Length

In real-life situations, columns are not singular members (as they have been portrayed in this chapter so far) but instead are elements of a structural frame. There are two basic categories of a structural frame: those that are **braced** against sidesway and those that are **unbraced** against sidesway. Frames can be braced against this lateral swaying in a number of ways, ranging from steel cross bracing to masonry shearwalls. However the question remains, in these real-life design situations, how is the effective length factor, K, estimated?

The method most commonly used to estimate the effective length factor, K, for these cases, known as the alignment chart method, develops a K value based on the restraint provided at the end of the column.

These alignment charts estimate the K factors based on the stiffness of all beams and columns that frame into the ends of the column. This essentially means that the end restraint of the column is largely due to the rigidity of the members at

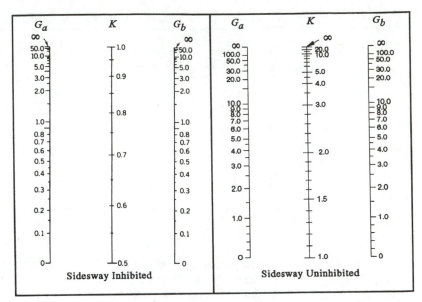

Figure 1. *The subscripts A and B refer to the joints at the two ends of the column section being considered. G is defined as*

$$G = \frac{\Sigma\,(I_c/L_c)}{\Sigma\,(I_g/L_g)}$$

in which Σ indicates a summation of all members rigidly connected to that joint and lying in the plane in which buckling of the column is being considered. I_c is the moment of inertia and L_c the unsupported length of a column section, and I_g is the moment of inertia and L_g the unsupported length of a girder or other restraining member. I_c and I_g are taken about axes perpendicular to the plane of buckling being considered.

For column ends supported by but not rigidly connected to a footing or foundation, G is theoretically infinity, but, unless actually designed as a true friction free pin, may be taken as "10" for practical designs. If the column end is rigidly attached to a properly designed footing, G may be taken as 1.0. Smaller values may be used if justified by analysis.

Figure 5–8 Standard Alignment Charts. (Courtesy of the American Institute of Steel Construction, Inc.)

that end of the column. The AISC has published an alignment chart for frames subjected to sidesway (sidesway uninhibited) and frames subjected to no sidesway (sidesway inhibited). The most widely used charts for sidesway inhibited behavior are the result of work by the Structural Stability Research Council.[4] The alignment charts for both cases are reproduced in Figure 5–8 with permission from the AISC.

The use of these alignment charts involves calculating a stiffness factor, G, at both ends of the column under consideration, using the following formula:

$$G_a = \Sigma(I_c/L_c)/\Sigma(I_g/L_g)$$

where G_a = stiffness at end a of the column.

The values I_c and I_g are the moment of inertias for the columns and girders, respectively, and L_c and L_g are their respective lengths. The summation must include only those members rigidly connected to that joint and lying in the plane for which buckling is being considered. The following example will demonstrate the basic use of these alignment charts in the calculation of the effective length factor, K. Again, when using these charts, please remember that buckling is taken as lying in the plane of the drawing and that the moments of inertia to be used for the columns and girders are taken perpendicular to this plane. Therefore, this generally means using the moments of inertia about the x–x or strong axis, unless buckling is to occur about the weak axis.

EXAMPLE 5.8

Calculate the effective length values for column 1 in the unbraced frame (sidesway uninhibited) shown in the figure. Assume columns are braced continuously in the weak direction.

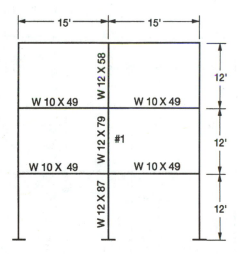

Let's begin by calculating the stiffness factor, G_a, at end A of the column in question. At joint A the sum of the (I/L)'s for all the columns at this end would be:

W 12 × 79 (the column itself) $I/L = 662$ in.^{4}/144 in. $= 4.60$

W 12 × 58 (the column above) $I/L = 475$ in.^{4}/144 in. $= 3.30$

Sum of columns I/L's $(\Sigma I/L)_{\text{columns}} = 4.60 + 3.30 = 7.90$

And the sum of the (I/L)'s for all the girders at joint A would be:

2–W 10 × 49's I/L = 272 in.^{4}/180 in. = 1.51

Sum of girders I/L's $(\Sigma I/L)_{\text{gird}}$ = 2 × 1.51 = 3.02

Therefore,

G_a = 7.9/3.02 = 2.62

Likewise, the stiffness at joint B can be found as follows:

For the columns, I/L for the W 12 × 79 = 4.60 and for the W 12 × 87 (the column below) I/L = 740 in.^{4}/144 in. = 5.14.

At joint B the sum of columns I/L's, $(\Sigma I/L)_{\text{col}}$ = 4.60 + 5.14 = 9.74.

Since the girders framing into joint B are of the same shape and length as at joint A, the sum of I/L's for the girders will be the same as before, 3.02.

Therefore, the stiffness at joint B, G_b, = 9.74/3.02 = 3.23. By entering these values for G_a and G_b in the alignment chart for sidesway uninhibited and connecting these values with a line (see the accompanying figure), the student can pick off an effective length factor, K, of ≈ 1.81.

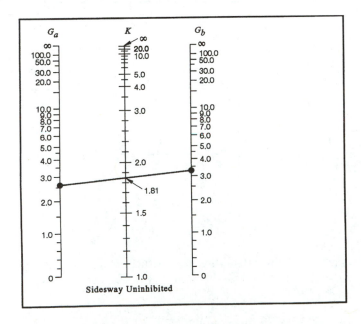

Another topic concerning the alignment charts is the stiffness reduction factors (SRF), found in Table A (p. 3–8) of the AISC manual and reproduced here in Figure 5–9. The use of these factors is to modify K values pulled off the alignment

Fy = 36 ksi		Table A						
Fy = 50 ksi		Stiffness Reduction Factors f_a/F_e'						

f_a	Fy 36 ksi	Fy 50 ksi	f_a	Fy 36 ksi	Fy 50 ksi	f_a	Fy 36 ksi	Fy 50 ksi
28.0	—	0.097	21.9	—	0.614	15.9	0.599	0.956
27.9	—	0.104	21.8	—	0.622	15.8	0.610	0.959
27.8	—	0.112	21.7	—	0.630	15.7	0.621	0.962
27.7	—	0.120	21.6	—	0.637	15.6	0.632	0.964
27.6	—	0.127	21.5	—	0.645	15.5	0.643	0.967
27.5	—	0.136	21.4	—	0.653	15.4	0.653	0.970
27.4	—	0.144	21.3	—	0.660	15.3	0.664	0.972
27.3	—	0.152	21.2	—	0.668	15.2	0.675	0.974
27.2	—	0.160	21.1	—	0.675	15.1	0.684	0.977
27.1	—	0.168	21.0	—	0.683	15.0	0.695	0.979
27.0	—	0.177	20.9	—	0.689	14.9	0.704	0.981
26.9	—	0.184	20.8	—	0.697	14.8	0.715	0.983
26.8	—	0.193	20.7	—	0.704	14.7	0.724	0.985
26.7	—	0.202	20.6	—	0.712	14.6	0.734	0.987
26.6	—	0.210	20.5	0.064	0.718	14.5	0.743	0.988
26.5	—	0.218	20.4	0.074	0.725	14.4	0.753	0.990
26.4	—	0.227	20.3	0.083	0.732	14.3	0.762	0.991
26.3	—	0.236	20.2	0.093	0.739	14.2	0.770	0.993
26.2	—	0.245	20.1	0.102	0.746	14.1	0.780	0.994
26.1	—	0.253	20.0	0.114	0.753	14.0	0.789	0.995
26.0	—	0.262	19.9	0.125	0.760	13.9	0.797	0.996
25.9	—	0.271	19.8	0.136	0.766	13.8	0.805	0.997
25.8	—	0.280	19.7	0.147	0.772	13.7	0.814	0.998
25.7	—	0.288	19.6	0.158	0.778	13.6	0.822	0.998
25.6	—	0.297	19.5	0.169	0.785	13.5	0.830	0.999
25.5	—	0.306	19.4	0.181	0.792	13.4	0.838	0.999
25.4	—	0.315	19.3	0.193	0.798	13.3	0.845	1.000
25.3	—	0.324	19.2	0.204	0.804	13.2	0.853	—
25.2	—	0.333	19.1	0.216	0.810	13.1	0.860	—
25.1	—	0.342	19.0	0.228	0.816	13.0	0.868	—
25.0	—	0.350	18.9	0.241	0.822	12.9	0.874	—
24.9	—	0.359	18.8	0.252	0.827	12.8	0.881	—
24.8	—	0.368	18.7	0.264	0.833	12.7	0.888	—
24.7	—	0.377	18.6	0.277	0.839	12.6	0.895	—
24.6	—	0.386	18.5	0.288	0.844	12.5	0.901	—
24.5	—	0.394	18.4	0.301	0.849	12.4	0.907	—
24.4	—	0.403	18.3	0.314	0.855	12.3	0.913	—
24.3	—	0.412	18.2	0.326	0.860	12.2	0.918	—
24.2	—	0.421	18.1	0.338	0.865	12.1	0.924	—
24.1	—	0.430	18.0	0.350	0.871	12.0	0.929	—
24.0	—	0.439	17.9	0.363	0.875	11.9	0.934	—
23.9	—	0.447	17.8	0.375	0.880	11.8	0.939	—
23.8	—	0.456	17.7	0.387	0.885	11.7	0.944	—
23.7	—	0.465	17.6	0.400	0.890	11.6	0.949	—
23.6	—	0.473	17.5	0.411	0.894	11.5	0.953	—
23.5	—	0.482	17.4	0.424	0.899	11.4	0.958	—
23.4	—	0.490	17.3	0.436	0.903	11.3	0.962	—
23.3	—	0.499	17.2	0.448	0.908	11.2	0.966	—
23.2	—	0.507	17.1	0.460	0.912	11.1	0.970	—
23.1	—	0.516	17.0	0.472	0.917	11.0	0.973	—
23.0	—	0.524	16.9	0.484	0.920	10.9	0.976	—
22.9	—	0.533	16.8	0.496	0.924	10.8	0.979	—
22.8	—	0.541	16.7	0.508	0.928	10.7	0.982	—
22.7	—	0.549	16.6	0.519	0.932	10.6	0.984	—
22.6	—	0.557	16.5	0.531	0.935	10.5	0.987	—
22.5	—	0.565	16.4	0.543	0.939	10.4	0.989	—
22.4	—	0.574	16.3	0.554	0.942	10.3	0.991	—
22.3	—	0.582	16.2	0.565	0.946	10.2	0.993	—
22.2	—	0.590	16.1	0.577	0.950	10.1	0.995	—
22.1	—	0.598	16.0	0.588	0.952	10.0	0.996	—
22.0	—	0.606				9.9	0.997	—
						9.8	0.998	—
						9.7	0.999	—
						9.6	1.000	—

Figure 5–9   Table of Stiffness Reduction Factors. (Courtesy of the American Institute of Steel Construction, Inc.)

charts if inelastic buckling controls. This modification is based on work by Joseph Yura[5] and is warranted because the alignment charts were developed for stresses in the elastic region. Since the inelastic region has a reduced modulus of elasticity, the K values should be likewise reduced.

The following example will demonstrate the utilization of both the alignment charts and the stiffness reduction factor.

EXAMPLE 5.9

For the unbraced frame (sidesway uninhibited) shown in the figure, calculate the adequacy of column 1. Calculate the effective length factor, K, from the alignment charts. The steel is A36.

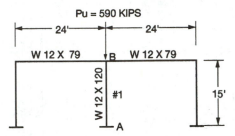

The first issue to determine in this problem is whether the column is in the elastic or inelastic range. This is done simply by calculating the actual stress of the column and comparing it to the chart in Figure 5–9. Because the actual stress, f_a, in this example is 590 kips/35.3 in.^{2} = 16.7 ksi, the column is controlled by inelastic behavior. The stiffness reduction factor from Figure 5–9 is $\approx$.508; therefore, all stiffnesses at the column ends should be reduced by this factor.

Next, calculate the stiffnesses at each end of the column. Since column end A is fixed in a footing, we will assume the stiffness G_a = 10 (per chart guidelines). Stiffness at end B is as follows:

Σ columns (W 12 × 120) = 1070 in.^{4}/15 ft × 12 in./ft = 5.94

Σ girders (2-W 12 × 79) = 662 in.^{4}/24ft × 12 in./ft = 2.30 × 2 = 4.60

Therefore G_b = 5.94/4.60 = 1.29, but this value should be reduced by a stiffness reduction factor of .508. Therefore, the actual stiffness at end B is:

G_b= 1.29 × .508 = .655

As mentioned before, the recommended value of G_a from the chart for unbraced frames is 10. The value of G_a is not reduced, because it is a value that was assumed to estimate the rigidity of the footing. The value is assumed to be 10 because that is a conservative estimate, and to maintain this conservatism we will leave it at 10. Finally, drawing a line (see the following figure) between .655 and 10 yields a K value of ≈1.84.

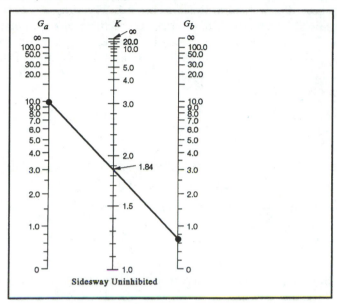

Solving for the allowable stress, F_a, is as follows:

Kl/r_x = 1.84(15 ft × 12 in./ft)/ 5.51 in. = 60.11.

Use AISC Eq. (E2–1) because 60.11 < 126.1 (C_c for A36)

Solving F_a = 17.42 ksi.

P_{all} = 17.42 ksi × 35.3 in.^{2} = 614.9 kips.

Since 614.9 kips > 590 kips, column works.

It should be noted that this example considered only in-plane behavior and x-axis bending. In real life, a designer may also have to consider buckling in the plane perpendicular to the paper if this direction might control.

5.10 Axially Loaded Column Baseplates

In a steel frame, columns typically transmit their loads to the column directly beneath them, such that the most heavily loaded columns are on the basement or bottom floor. At this stage, the column will then transmit its load into a footing or foundation system that is commonly made of concrete or masonry. Since the allowable compressive strength of concrete (and masonry) is much smaller than that of steel, the load from the column must be spread over a larger area of the foundation. This is accomplished through the use of a steel baseplate that is typically bolted or welded to the bottom of the column.

The allowable stress that the AISC specification recommends for the concrete bearing surface depends on the area of the bearing plate (referred to as A_1) and the area of the concrete support (referred to as A_2). If the area of the bearing plate, A_1, is equal to the area of concrete support, A_2, the allowable bearing stress on the concrete is given as follows:

$$F_p = .35 f_c'$$

where

f_c' = specified concrete strength

F_p = allowable concrete bearing stress

Should the concrete support area be larger than the baseplate area (which is very typical), the allowable bearing stress is given as follows:

$$F_p = .35 f_c' \sqrt{(A_2/A_1)} \leq .7 f_c'$$

Since typical footings have relatively large support areas, many times the value of allowable bearing stress becomes $.7 f_c'$.

The area of the baseplate, A_1, can be found simply by rearranging the previous allowable stress equation and solving for the larger of the two baseplate areas as follows:

$$A_1 = (P/.35 f_c')^2/A_2$$

or

$$A_1 = P/.7 f_c'$$

where P is the axial load on the column baseplate.

The final part of axial baseplate design involves the selection of the appropriate thickness of the plate itself. This thickness is based on the bending behavior of the plate caused by moments that tend to "curl up" the plate along its cantilevered edges (Figure 5–10). Based on this behavior, formulas have been developed to calculate plate thickness. The values referred to as m and n are the cantilevered distances of the baseplates that are subjected to the maximum bending stresses. These distances are diagrammatically shown in Figure 5–11. The formulas to calculate plate thickness are as follows:

$$t = 2n\sqrt{f_p/F_y} \quad \text{and} \quad t = 2m\sqrt{f_p/F_y}$$

where

f_p = actual bearing stress

F_y = yield stress of baseplate steel

The following example will illustrate the use of these equations in a typical baseplate design.

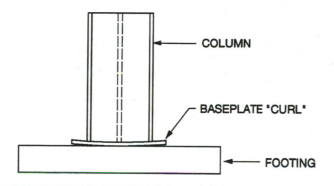

Figure 5–10   Effect of Bearing Pressures on Column Baseplates.

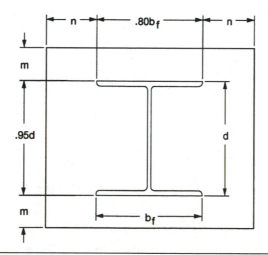

Figure 5–11   Standard Dimensions for Column Baseplates.

EXAMPLE 5.10

Design a square baseplate made from A36 steel for a W 12 × 65 column that rests on a 3 ft × 3 ft concrete footing. The column is under an axial load of 200 kips, and the concrete has an f'_c of 3 ksi.

To begin, let's get the size of the steel baseplate that will be required, by choosing the larger of the two previously mentioned formulas.

$$A_1 = [200 \text{ kips}/.35(3 \text{ ksi})]^2/(36 \text{ in.} \times 36 \text{ in.})$$
$$= 28.0 \text{ in.}^2$$

or

$$A_1 = 200 \text{ kips}/.7(3 \text{ ksi}) = 95.2 \text{ in.}^2$$

Choosing the larger area, we may be tempted to design a 10 in. × 10 in. square plate, but realizing that a W 12 × 65 is 12.12 in. deep and has a flange width of 12 in., it is best to select a 14 in. × 14 in. plate to provide adequate room for a proper connection.

Next we can solve for the thickness of the plate by calculating the previously mentioned equations. But first we must calculate the cantilevered distances, m and n. As shown in the accompanying figure, m is found to be equal to 1.24 in. and n is equal to 2 in. Now we can calculate the actual bearing stress as

$$f_p = 200 \text{ kips}/14 \text{ in.} \times 14 \text{ in.} = 1.02 \text{ ksi}$$

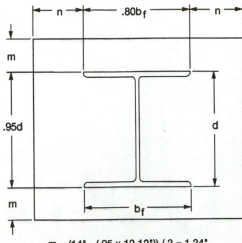

$$m = (14" - (.95 \times 12.12")) / 2 = 1.24"$$
$$n = (14" - (.80 \times 12.00")) / 2 = 2.2"$$

Now, solving for the thickness we find:

$$t = 2(1.24)\sqrt{1.02\,\text{ksi}/36\,\text{ksi}} = .42\,\text{in.}$$

and

$$t = 2(1.0)\sqrt{1.02\,\text{ksi}/36\,\text{ksi}} = .34\,\text{in.}$$

Using the larger thickness (.42 in.), our plate probably will be either 7/16 in. or 1/2 in. thick.

5.11 Singly Symmetrical and Unsymmetrical Members in Compression

Up to now, the discussion of compression members has only dealt with the doubly symmetrical shape of the wide-flange section. This type of section is rather nice because it typically fails by flexural buckling and is not prone to torsional-related failures. The calculation of torsional resistance for compression members is a long and complicated procedure that is beyond the scope of this text. Unfortunately, other singly symmetrical or unsymmetrical shapes are sometimes used in compressive situations. Such shapes may include angles, double angles, structural tees, and builtup sections. Singly symmetrical sections, such as the double angle, and unsymmetrical sections, such as angles, are very susceptible to a failure known as **flexural-torsional buckling**.

Flexural-torsional buckling is a very complicated behavior, and it is aggravated by the fact that loading eccentricities among such unsymmetrical shapes are very prevalent. Therefore, if the singly symmetrical or unsymmetrical shapes are to be used as compressive elements, both flexural buckling and flexural-torsional buckling must be checked for proper adequacy. If the reader should encounter such situations, there are many advanced design texts that address this topic.

5.12 Summary

Columns, and compression members in general, are some of the most important elements in the scope of structural design. Failure of such members primarily occur due to an out-of-plane bending referred to as buckling. Buckling of a column is very dependent on the length of the member, which is quantified in design by the concept of slenderness ratio. Another important element contained in the slenderness ratio is the effective length factor, K. This factor takes into account the column's end-restraint effect and is used to estimate the actual buckling length.

The AISC specification considers two potential types of buckling behavior for steel columns: elastic and inelastic. When columns are shorter (and less slender), they will tend to fail due to inelastic buckling and the AISC specification will therefore use a larger allowable stress. Conversely, as a column becomes longer, it will be more likely to fail due to elastic buckling and proper use of the specifications will therefore lead to a smaller allowable stress.

EXERCISES

1. Using AISC Eqs. (E2–1) and (E2–2), construct a graph for the curve that describes the AISC philosophy of column behavior. Use slenderness ratios from 10 to 180 in increments of 10.

2. Why are the recommended values on the effective length factor for idealized conditions always equal to or higher than the theoretical values? Is this practice conservative?

3. How does the presence of residual stresses affect the buckling strength of columns? How do accidental eccentricities or crookedness of a column affect the buckling strength?

4. Given an unbraced W 10×49 column 20 ft long, pinned at both ends, calculate the allowable load, P_{all}. The steel is A242 ($F_y = 50,000$).

5. An unbraced W 12×79 is 15 ft long and carries a load of 149 kips. If the $K = 1.0$ and the column is made from A36 steel, determine whether it is adequate.

6. Design the most economical W12 section to carry a concentric axial load of 375 kips. The column is pin–fix at its ends and is unbraced. The steel is A36 and the column is 22 ft long.

7. Design the most economical W10 section to hold a load of 240 kips, if it is 13 ft long and unbraced with a $K = 1.0$. The steel is A242 ($F_y = 50$ ksi).

8. Rework Exercise 7, this time assuming that the column is braced at midheight in the weak axis.

9. Calculate the allowable load, P_{all}, on a W 12×40 column that is 10 ft long with pinned end conditions, if it is braced at midheight in the strong direction. The steel is A36.

10. In the unbraced frame shown in the accompanying figure, calculate the effective length factors for columns 1, 2, and 3 using the appropriate alignment charts. Using this information, calculate the maximum allowable load that each column can carry. Remember to include stiffness reduction factors where applicable. The load on all columns is 300 kips, and the steel is A36. Assume continuous out-of-plane bracing in the weak direction.

11. Repeat Exercise 9, assuming the column height is increased to 11 ft.
12. Rework Exercise 10, assuming sidesway is inhibited due to lateral bracing. How does this increase your design strength of these columns? Calculate the percentage increase.
13. Calculate the allowable capacity of a 4 × 4 × 1/2 steel tube whose ends are assumed to be pinned. The column is 20 ft long and is unbraced. Use steel with F_y = 46 ksi.
14. Design an unbraced column from a rectangular steel tube to hold an axial load of 300 kips, if the column is 15 ft long with pinned end conditions. Use steel with F_y = 46 ksi.
15. The columns in the frame shown in the accompanying figure are part of a typical 15 ft × 20 ft bay in a proposed warehouse. Assuming an equal load distribution based on distributive area and neglecting bending induced from the beams, design the lightest W-section to hold the following loads. Steel is A36 and use K = 1.0.
    a. Self-weight of all members
    b. Live load of 500 psf plus dead load

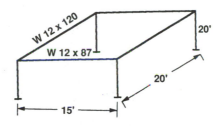

16. Design a square baseplate under a W 12 × 79 column carrying an axial load of 275 kips. The baseplate rests on a 4 ft × 4 ft concrete footing with an f'_c = 4000 psi.

REFERENCES

1. L. Euler, with English translation by J.A. Van den Broek, "Euler's Classic Paper 'On the Strength of Columns,'" *American Journal of Physics,* January 1947, pp. 309–318.
2. Jack McCormac, *Structural Steel Design* (New York: Harper & Row, 1981), p. 84.
3. Patrick Dowling, Peter Knowles, and Graham Owens, *Structural Steel Design* (London: The Steel Construction Institute, 1988), p. 156.
4. Bruce G. Johnston, ed., Structural Stability Research Council, *Guide to Stabil-*

ity Design Criteria for Metal Structures, 3rd ed. (New York: Wiley and Sons, 1976).

5. Joseph A. Yura, "The Effective Length of Columns in Unbraced Frames," *Engineering Journal,* American Institute of Steel Construction, April 1971, pp. 37–42.

6

ROLLED-BEAM DESIGN

6.1   Introduction and Review of Beam Theory

In this chapter we will attempt to explore the fundamental concepts behind the ASD philosophy concerning the design of rolled-steel beams. Although some may believe that limiting the discussion to rolled-steel beams offers incomplete coverage to the topic of beam design, the author contends that a firm grasp of the fundamentals is needed before any discussion can occur on more complex design topics such as plate girders and composite beams. As more advanced topics such as these are explored, the student will find that a solid knowledge of rolled-beam design is extremely helpful.

Beams, girders, stringers, purlins, and joists are all terms that describe members that may support loads applied perpendicular to their longitudinal axis. These members are integral pieces of all structures, from the small steel beam found in the basement of typical residences to the large steel girders supporting interstate overpasses (Figure 6–1).

The stresses involved in the bending of beams should have been covered thoroughly in a course on strength of materials. However, a brief review of the basic concepts will follow in the next few paragraphs.

When a beam is subjected to applied loadings, it has to develop two primary stresses in order to maintain its integrity. To keep from ripping apart, the beam develops shear stresses to resist the shear resulting from the applied loads. Further,

Figure 6–1   Steel Girder Construction on the Passaic River Bridge, New Jersey. (Courtesy Bethlehem Steel Corporation.)

to keep from deflecting or rotating excessively, the beam must develop internal bending stresses to resist the applied bending moments caused by the loads (Figure 6–2). If the beam cannot develop resisting stresses to offset the stresses resulting from the applied loading, failure will occur. A brief review of these stresses follows.

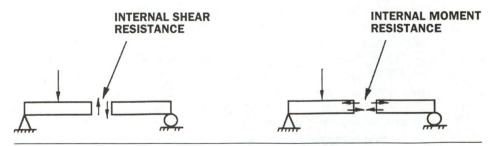

Figure 6–2   Internal Beam Resistance to Shear and Bending Forces.

Shear Stresses

The actual shear stress, f_v, developed by a beam has to be equal to the shear stress applied to the beam. This shear stress is typically calculated from the formula

$$f_v = \frac{VQ}{Ib}$$

where

V = shear force at point under consideration, kips or lb

Q = static moment of area, in.[3]

I = moment of inertia of beam, in.[4]

b = width of beam at point under consideration, in.

The reader should remember that the shear stress at the neutral axis of a beam will always be the maximum value. This fact is further magnified on a steel beam section, since the width at the neutral axis is only the width of the web. Example 6.1 will refresh your memory on the use of this formula.

EXAMPLE 6.1

Calculate the shear stresses at the neutral axis and at the midpoint of the flange in a W 16 × 77 if the shear force is 100 kips.

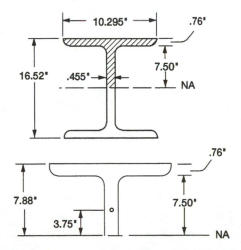

In this problem the majority of work will go into the calculation of the static moment of area, Q, since moment of inertia, I_x, and the width of the web and the flange can readily be taken from the appropriate section table. Remember that Q is simply the area outside the plane under consideration multiplied by the distance from the neutral axis to the centroid of that area.

For shear stress at the neutral axis, $f_{v@N.A.}$, the Q can be calculated by breaking down the area outside the neutral axis into two rectangles. The Q can then be readily found by summing the areas multiplied by their respective distances (from N.A. to their centroids; see diagram):

$$Q_{@N.A.} = (7.5 \text{ in.} \times .455 \text{ in.})(7.5 \text{ in.}/2) + (10.295 \text{ in.} \times .76 \text{ in.})(7.88 \text{ in.})$$
$$= 74.45 \text{ in.}^3$$

$$Q_{@midflange} = (10.295 \text{ in.} \times .38 \text{ in.})(8.07) = 31.57 \text{ in.}^3$$

Therefore the shear stress can be simply calculated as follows:

$$f_{v@N.A.} = 100 \text{ kips} \times 74.4 \text{ in.}^3/1110 \text{ in.}^4 \times .455 \text{ in.} = 14.73 \text{ ksi}$$

$$f_{v@midflange} = 100 \text{ kips} \times 31.57 \text{ in.}^3/1110 \text{ in.}^4 \times 10.295 \text{ in.} = .27 \text{ ksi}$$

We can see that the shear stresses at the neutral axis are much greater than other parts of the section. It should be pointed out that the shear stress at the junction of the flange and web is much higher than at the midpoint of the flange. (The author would also note that the fillets at the junction of the flange and web were excluded to simplify this example. If the student has access to the WT 8 × 38.5 section properties, the $Q_{@N.A.}$ can be calculated as 74.92 in.^{3} with the fillets included.)

Flexural Stress

The stress that the beam develops in response to applied bending moment is termed **flexural stress** or **bending stress**. The formula to calculate actual bending stress, f_b, is as follows:

$$f_b = \frac{Mc}{I}$$

where

M = applied moment at point under consideration (usually the maximum moment along the beam), kip-ft or kip-in.

c = distance from neutral axis to point on beam cross section desired (usually outer fiber for design), in.

I = moment of inertia about bending axis, in.^{4}

The student should be aware that because designers are typically concerned with maximum stress, the value of c will usually be 1/2 the beam depth. This will

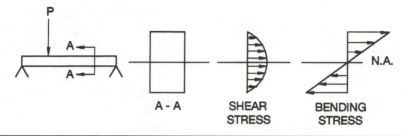

Figure 6–3   Stress Distribution over a Rectangular Beam Cross Section.

be the case on wide-flange sections, since they are symmetric sections with the neutral axis located at middepth. Bending stress is assumed to vary proportionally with the distance c in a homogeneous member and therefore reaches a maximum value at the extreme outside (top and bottom) of a section.

The stress distributions over a rectangular beam can be seen in Figure 6–3. Again, it should be reiterated that shear is maximum at the neutral axis, while the bending stress is zero at this point. Conversely, bending stress is maximum at the outer fiber of the beam, while shear stress is zero.

A common revision of the flexural stress formula in beam design is the substitution of the elastic section modulus, S, for the ratio of I/c. The flexural stress formula can then be rewritten as

$$f_b = \frac{M}{S_x}$$

where S_x is the section modulus of the beam (equal to I/c) usually taken about the strong or x-axis, in.^{3}.

Example 6.2 will illustrate the application of the bending stress formula and should be a helpful review.

EXAMPLE 6.2

Determine the bending stress at the outside fiber for a W 12 × 79 under a moment of 120 kip-ft.

The moment of inertia (I_x) for this section is listed in the AISC tables (pp. 1–28 and 1–29) as 662 in.^{4}, and the distance away from the neutral axis is 1/2 the depth, or 12.38 in./2 = 6.19 in. The student can conversely use the tabulated value of S_x = 107 in.^{3}, since this is simply the ratio of I/c (662 in.^{4}/6.19 in.).

The calculation of bending stress is as follows, but notice that the moment is converted into kip-inches for unit compatibility.

f_b = 120 kip-ft × 12 in./ft ÷ 107 in.^{3} = 13.46 ksi

The section modulus, S_x, is an elastic property of area that is useful only as stresses are kept in the elastic region of the material, that is, under the proportional limit. In the allowable stress philosophy, the section modulus is exclusively used in design, since one failure mode can be considered to occur at the first yield of any fiber on the cross section. Although failure may be thought to occur at first yield, it is well known that a beam has reserve capacity far exceeding the moment causing this first yield.

This ability to develop a higher capacity, while *all* fibers throughout the cross section are yielding, depends upon the individual elements of that beam remaining stable. A beam that has this stability while all fibers reach yield stress is referred to as being **compact**. The moment capacity attainable at this state of full yield is referred to as the **plastic moment**.

Through research,[1] it is known that rolled-beams have a plastic moment capacity of at least 10%–12% greater than the moment at first yield. When a beam reaches this plastic moment capacity, it has yielded every fiber along its cross section and is said to have developed a "plastic hinge" (Figure 6–4). The term *plastic hinge* essentially reflects the fact that the beam has no more rotational capacity at this point and will most probably collapse if no redistribution of moment occurs.

6.2 Potential Modes of Beam Failure

The behavior of a beam is a mixture of the two types of members that we have discussed up to this point: tension and compression members. Therefore, it only stands to reason that the types of failure that a beam can undergo are essentially a mixture of the failure modes that we have studied in previous chapters.

The tension side of a beam experiences tensile stresses that are highest at the extreme outside fiber. The mode of failure we are concerned about on the tension

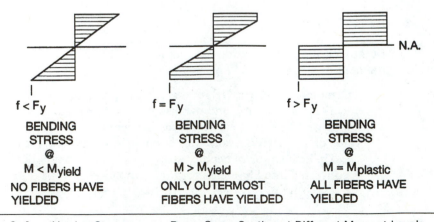

Figure 6–4 Varying Stresses over Beam Cross Section at Different Moment Levels.

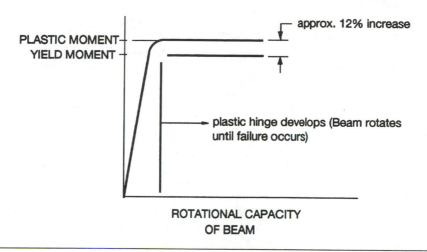

Figure 6–5   Rotational Behavior of a Steel Beam.

side of the beam would involve excessive deformation, due to the reaching and exceeding of the yield stress, F_y. Because the stresses and strains over the beam's cross section are assumed to vary linearly, the compression side of the beam would also be subject to the same stresses. However, the major concern on the compression side of the beam is its ability to remain stable under these stresses. Failure can occur due to instability on the compression side of the beam in two potential manners: local buckling and lateral torsional buckling.

Local buckling is an instability by which the individual elements of a beam (the flanges and the web) may fail under compressive stress before the beam is able to reach its plastic moment capacity. Remember that the ability of a steel shape to reach yield stress across its entire cross-sectional area before failure is referred to as compactness. If a beam is compact, its individual elements will remain stable and exhibit no local buckling until the section reaches its plastic moment capacity. This capacity is attained when all the fibers have yielded and is, on average, 10%–12% greater than the moment needed to produce first yield.[2] (See Figure 6–5.)

The AISC criteria for determining compactness, found in Table B5–1 (page 5–36) of the AISC manual, basically compares the width/thickness ratio of the compression elements to some limiting value. The applicable flange and web compactness criteria have been reproduced in Table 6–1. Notice that there are two levels in which the compactness of a beam can be measured: compact and noncompact. These criteria can simply be thought of as dividing lines assuring that the beam will reach its full plastic capacity (be compact) or at least will behave inelastically (be noncompact). It should be noted that practically all rolled-steel shapes meet the compactness criteria and therefore can achieve full plastic capacity.

The other instability affecting the development of a beam's resistance to bend-

Table 6–1 AISC Compactness Criteria for Common Rolled-Beam Elements

Element Description	Width/Thickness Ratio	Limiting Ratios (b/t)	
		Compact	Noncompact
Flanges of rolled beams	$b_f/2t_f$	$65/F_y$	$95/F_y$
Webs in flexural compression	d/tw	$640/F_y$	—
Unstiffened elements supported on one side such as single angle struts	d/tw	NA	$76/F_y$

ing stress reflects the condition of lateral support along its compressive flange. This column-type instability in beams is referred to as **lateral torsional buckling,** because as the slender compression flange begins to buckle out of plane, the beam undergoes a torsional component due to the downward forces along the top flange (Figure 6–6).

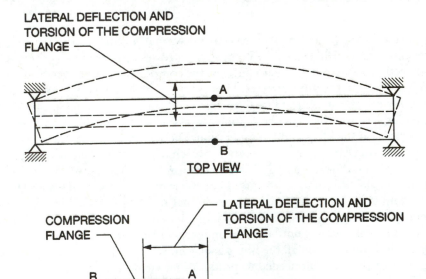

Figure 6–6 Illustration of Lateral Torsional Buckling Behavior.

As discussed in the preceding chapter on columns, bracing of the buckling plane will make it more difficult for failure to occur by buckling. Bracing the compression flange will increase the beam's ability to resist stresses.

A beam can have many conditions of lateral support along its compression flange. It can be fully supported (braced everywhere along its flange), partially supported (braced at intermittent points along its flange), or unsupported (having no external support along its flange). Methods that achieve a degree of lateral support include the following:

- Noncomposite flooring (at least by friction)
- Composite flooring
- Cross bracing
- Beams or struts that frame in at certain points

The assumption of what type of lateral support exists in a certain situation (unsupported, partial, full) is a judgment left to the designer. It should be noted that rarely is a beam 100% unsupported along its compression flange, since most beams would be subject to frictional restraint caused by the floor system's self-weight.

6.3   Rolled-Beam Behavior

From our discussion in the previous section, it can be seen that beam behavior is largely a function of compactness and support of the compression flange. The following discussion will try to relate beam behavior as a function of these two criteria. Remember that most rolled-steel beams will meet all compactness requirements; therefore, the majority of our discussion will focus on the lateral bracing criteria concerning compact members. Some effort will also be spent on noncompact member behavior.

In a compact rolled beam being bent about the strong axis, the controlling criterion affecting capacity of the beam is the support of the compression flange. The more "fully" supported the compression flange is, the more likely it is that failure will occur by plastic hinge formation and the less likely it is that lateral torsional buckling will occur. Conversely, the farther apart the support along the compression flange is spaced, the more likely it is that the beam will fail due to lateral torsional buckling prior to reaching its full plastic capacity. Lateral torsional buckling is very similar to column behavior, in that it incurs two types of possible buckling: inelastic and elastic. Inelastic lateral torsional buckling takes place when the distance between braced points is short enough that it allows the beam to develop higher stresses that happen to be in the material's inelastic region. Elastic lateral torsional buckling occurs when the distance between lateral supports of the compression flange becomes large so that the stresses developed in the beam at failure are relatively low—below the material's elastic limit. (Is the similarity between column behavior and lateral torsional buckling evident?) We can best understand this concept if we imagine a beam of infinite length that is initially supported at every point along its compression flange. This fully supported beam cannot fail due to lateral torsional buckling, but if these points of support are taken away, one by one, at some point the beam will fail due to buckling. This beam initially would be able to reach its full plastic capacity, but this capacity would be diminished as the lateral support slowly erodes away.

To determine whether a rolled-steel beam is susceptible to lateral torsional buckling, the AISC compares the beam's actual unbraced length, termed l_b, to a limiting length based on the section known as L_c. If the beam's actual unbraced length, l_b, is less than its L_c value, lateral torsional buckling will not occur.

The value of L_c is the *smaller* of two calculated equations based on a section's resistance to lateral torsional behavior. These two equations for L_c are as follows, the first approximating the length at which column buckling of the compression flange will control failure and the second approximating the length at which torsional resistance will control relative to the section's plastic capacity:

$$L_c = \frac{76\,b_f}{\sqrt{F_y}} \quad \text{or} \quad \frac{20,000}{(d/A_f)F_y}$$

If the beam's unbraced length, l_b, exceeds the smaller of these two equations ($l_b > L_c$), lateral torsional buckling will take place prior to the section's achieving its plastic capacity.

Compact behavior in rolled beams, as was previously mentioned, is not a great concern in the vast majority of rolled-beam shapes, because most shapes meet the proposed criteria. The AISC recognizes that a section must be compact to be able to reach its plastic capacity. Therefore, compactness criteria are relevant only when lateral torsional buckling behavior is also prevented, or when $l_b \leq L_c$.

Some sections may exhibit what is known as "noncompact" behavior, especially when higher-strength steels are used. This occurs when an individual element of the section (most likely the flange) exceeds the compactness limit but is still less than the noncompactness limit, as set forth previously in Figure 6–6. Again, this behavior is relevant only when the lateral torsional mode of failure is prevented, when $l_b \leq L_c$. This topic will be discussed further in the following section.

6.4 AISC Rolled-Beam Philosophy Using ASD

The AISC specification (in Chapter F of its manual) sets the criteria for determining allowable bending stress, F_b. The allowable bending stress determination for rolled-steel beams and channels can be thought of as falling in one of the three following categories:

Category 1: Members with compact sections and $l_b \leq L_c$

Category 2: Member with noncompact sections and $l_b \leq L_c$

Category 3: Compact or noncompact sections with $l_b > L_c$

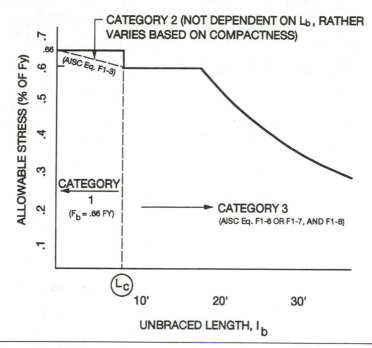

Figure 6–7   Allowable Bending Stress Categories for an Ideal Beam.

These categories will be discussed in the upcoming pages and should be thought of as a hierarchy of allowable bending stress levels ranging from the best (category 1) to the worst (category 3). The AISC philosophy governing rolled beams is graphically represented in Figure 6–7; students are encouraged to refer to this figure often as they read about these categories.

Category 1: Compact Sections and $l_b \le L_c$

Beams in this category are able to achieve their plastic capacity by preventing local buckling (being compact) and lateral torsional buckling (having $l_b \le L_c$). Since failure due to these two behaviors is prevented, this category represents the best possible type of beam behavior. The AISC rewards beams falling in this category by setting the allowable bending stress level at 0.66 F_y. This is the highest allowable bending stress level awarded to beams bent about their strong axis.

The lateral torsional buckling criteria are met by ensuring that the beam's unbraced length, l_b, is less than or equal to the *smaller* of the two L_c values, which were discussed in the previous section.

The compactness criteria were found earlier in Table 6–1. Remember that the flange value, $b_f/2t_f$, and the web value, d/t_w, are checked against the appropriate criteria.

Category 2: Noncompact Sections and $l_b \leq L_c$

Beams that meet the unbraced length requirement ($l_b \leq L_c$) but fail to meet flange compactness criteria will fall into this category. Recognizing that local buckling will not let the section reach its full plastic capacity, the AISC provides an equation setting the allowable stress at a transitional value between 0.66 F_y and 0.60 F_y. This equation is as follows:

$$F_b = F_y \times \left[0.79 - .002 \left(\frac{b_f}{2t_f} \right) \sqrt{F_y} \right] \quad \text{(AISC Eq. F1–3)}$$

If a beam should fail both the web and flange criteria regarding compactness but still meet the unbraced length criteria, the AISC sets the allowable bending stress to 0.60 F_y.

Category 3: Any Beam (Compact or Noncompact) with $l_b > L_c$

The primary mode of failure for beams in this category is by lateral torsional buckling, since the unbraced length criterion ($l_b > L_c$) is exceeded. The compactness of the rolled-beam section is irrelevant since the lateral torsional behavior is dominant. The allowable bending stress that the AISC specifies is the larger calculated from two equations, **but in no case can the allowable stress be higher than 0.60 F_y for beams in this category.**

The two equations that will be used consist of a torsional equation and a lateral buckling equation of the compression flange. However, as with column buckling, when the lateral buckling behavior is considered, we will have to decide whether elastic or inelastic behavior controls. As the unbraced length of the compression flange gets longer, elastic behavior will control; and as the unbraced length gets smaller, inelastic behavior will control. The slenderness ratio of the compression flange, termed l/r_T, is compared to certain AISC dividing lines to determine whether the lateral behavior is elastic or inelastic. (In the slenderness ratio of the compression flange, l is equal to the unbraced length and r_T is the radius of gyration of the compression flange, which can easily be found in the section tables. It should be noted that the "compression flange" actually is considered to also encompass a small part of the web). If elastic behavior controls, AISC Eq. (F1–7) (page 5–47) is calculated, but if inelastic behavior controls, AISC Eq. (F1–6) is calculated. The torsional equation is AISC Eq. (F1–8) and is always calculated. After calculating **either** AISC Eq. (F1–6) or (F1–7) **and** AISC Eq. (F1–8), the larger value calculated is used for the allowable stress. However, it is important to remember that in no case can the allowable bending stress exceed 0.60 F_y in category 3. The AISC equations for allowable stress are as follows:

When $\sqrt{\dfrac{102 \times 10^3 C_b}{F_y}} \leq l/r_T \leq \sqrt{\dfrac{510 \times 10^3 C_b}{F_y}}$

$$F_b = \left[\frac{2}{3} - \frac{F_y \left(l/r_T \right)^2}{1530 \times 10^3 C_b} \right] F_y \qquad \text{(inelastic buckling, AISC Eq. F1 – 6)}$$

When $l/r_T \geq \sqrt{\dfrac{510 \times 10^3 C_b}{F_y}}$

$$F_b = \frac{170 \times 10^3 C_b}{\left(l/r_T \right)^2} \qquad \text{(elastic buckling, AISC Eq. F1 – 7)}$$

In every case, calculate

$$F_b = \frac{12 \times 10^3 C_b}{l_d / A_f} \qquad \text{(torsion strength, AISC Eq. F1 – 8)}$$

The C_b term is a modifier used to define the effect of moment gradient along the beam. It will be discussed further in Section 6.5 of this chapter; for purposes of simplification, at this point in time the value of C_b will be taken as 1.0.

The following examples will demonstrate the use of these allowable bending stress formulas in some typical evaluation-type problems.

EXAMPLE 6.3

Determine the adequacy of a W 24 × 117 beam that has full lateral support and is a simply supported 10-ft span loaded with a uniform load of 12 kips per foot. The steel is A36.

First let's determine the allowable bending stress per AISC specification criteria. Since the beam is said to have full lateral support, the unbraced length, l_b, is zero and less than any calculated value of L_c. Therefore, if the section meets compactness criteria, it will fall in category 1.

Check the section for compactness:

$b_f / 2t_f = 7.5 \leq 65 / \sqrt{36} = 10.8$ Flanges are compact.

$d/t_w = 44.1 \leq 640 / \sqrt{36} = 106.6$ Web is compact.

Since the section is compact and fully supported against lateral buckling (as well as symmetrical and bent about its strong axis), the allowable bending stress falls in category 1.

The allowable bending stress, F_b = 0.66 (36 ksi) = 23.76 ksi. (*Note:* Many designers prefer to round 23.76 ksi to 24 ksi.) Now we must find the actual stress caused by the uniform load of 12 kips per foot over this beam. We can find this moment by drawing shear and moment diagrams or by the following equation for maximum moment on a uniformly loaded beam:

$$M = wl^2/8 = (12 \ k/ft)(10 \ ft)^2/8 = 150 \ kip\text{-}ft$$

Calculating actual bending stress using this maximum moment yields:

$$f_b = M/S = 150 \ kip\text{-}ft \times 12 \ in./ft \div 291 \ in.^3 = 6.19 \ ksi$$

Since the actual bending stress (6.19 ksi) is less than or equal to the allowable bending stress (23.76 ksi), the beam is adequate in bending. (It should be noted that the compactness values for $b_f/2t_f$ and d/t_w are already precalculated in the AISC section tables.)

EXAMPLE 6.4

Determine the adequacy of the W 12 × 87 shown in the figure, if the compression flange is unbraced. The steel is A36.

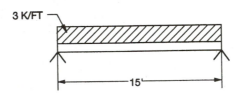

Again, we will begin by trying to determine the value of allowable bending stress for this section under the stated conditions. Being unsupported means that the unbraced length, L_b, is 15 ft or 180 in. Before checking the compactness values listed for a W 12 × 87 (since we won't need them if this falls in category 3), let's calculate the values for L_c:

$$L_c = 76b_f \, / \sqrt{F_y} \qquad \text{or} \qquad \frac{20000}{(d/A_f)F_y} \qquad \text{(the smaller of)}$$

L_c = 12.7 ft; therefore, $l_b > L_c$ and the section falls into category 3 for determination of allowable bending stress. In category 3 the allowable stress is the larger of the calculated stresses from Eq. (F1–6) **or** (F1–7) **and** Eq. (F1–8). But in no case can it ever be greater than 0.60 F_y.

First, determine whether to use Eq. (F1–6) or (F1–7). This is based on the slenderness ratio of the compression flange, l/r_T. Calculate l/r_T:

$$\frac{l}{r_T} = \frac{15 \times 12}{3.32} = 54.2$$

With $C_b = 1.0$ (simple support), our l/r_T value falls between the two AISC dividing lines for inelastic lateral buckling, as follows:

$$\sqrt{102,000(1.0)/36} = 53.2 \leq l/r_T \leq \sqrt{510,000(1.0)/36} = 119.0$$

Therefore, calculate AISC Eqs. (F1–6) and (F1–8) and use the larger.
For AISC Eq. (F1–6),

$$F_b = (2/3 - [36(54.2)^2/(1,530,000(1.0))]) \times 36 \text{ ksi} = 21.52 \text{ ksi} \leq 21.6 \text{ ksi}$$

For AISC Eq. (F1–8),

$$F_b = 12,000(1.0)/[180(1.28)] = 52.08 \text{ ksi not less than } 0.60 \ F_y$$

Therefore, use $0.60 \ F_y = 21.6$ ksi for AISC Eq. (F1–8). *Larger of the two AISC equations is 21.6 ksi.*
Now calculate the actual maximum moment placed on this beam, as we did in the preceding example:

$$M = (3 \text{ kips/ft})(15 \text{ ft})^2/8 = 84.38 \text{ kip-ft}$$

Calculating the actual bending stress caused on the section by this moment yields

$$f_b = M/S = 84.38 \text{ kip-ft} \times 12 \text{ in./ft} \div 118 \text{ in.}^3 = 8.58 \text{ ksi}$$

Since the actual is less than the allowable stress, the beam is adequate in bending.

EXAMPLE 6.5

Calculate the maximum allowable moment that can be applied to a W 14 × 90 section that is fully supported and made from A242 steel ($F_y = 50$ ksi).

First, let's calculate the allowable bending stress associated with this section. Since this beam is fully supported, the unbraced length is zero and it will fall in category 1 if it meets compactness criteria. Checking compactness values as found in the AISC section tables against the compactness limits, we find:

$$b_f/2t_f = 10.2 > 65/\sqrt{50} = 9.2 \quad \text{Flanges do not meet criteria, but they are noncompact since } 10.2 \leq 95/\sqrt{50}.$$

$$d/t_w = 31.9 \leq 640/\sqrt{50} = 90.5 \quad \text{Web is compact.}$$

Therefore, the allowable stress falls in category 2 and is calculated from the AISC equation (F1–3) as follows:

$$F_b = 50 \text{ ksi} \times \left[0.79 - 0.002(10.2)\left(\sqrt{50}\right)\right] = 32.29 \text{ ksi}$$

Now calculate the maximum allowable moment on this section as follows:

$$32.29 \text{ ksi} \times 143 \text{ in.}^3 = 4617.5 \text{ kip-in. or } 384.79 \text{ kip-ft}$$

6.5   The Bending Coefficient, C_b

Up to this point, you may have noticed the modifier, C_b, in the moment capacity equations the AISC has presented. Why is this modifier needed, and what does it actually do to the equations we use in rolled-steel beam design?

In the development of its design equations, the AISC realized that the worst possible case for causing lateral torsional buckling would be to have the compression flange under constant uniform moment over its unbraced length. This worst case was therefore used as a conservative basis for the formulation of the AISC Eqs. (F1–6, F1–7, F1–8), which we presently use. But realizing that as the moment varied over the unbraced length, the failure mode of lateral torsional buckling was less likely to occur, the AISC introduced C_b as a modifier to account for this behavior. We might understand this concept more easily by envisioning a beam being broken up into individual unbraced segments. The segment that is most likely to undergo lateral torsional buckling (let's call this segment MAX) would be the segment under the highest applied moment over the beam length, because this moment will induce the largest compressive forces into the flange. The segments adjacent to segment MAX are less likely to undergo buckling, since the total compressive force is smaller and the moment gradient may also have been changed. In fact, because the segments will be less likely to buckle laterally, they will actually provide a restraint on segment MAX as it tries to undergo lateral torsional buckling. This restraining effect can be viewed similarly to the effective length factor, K, which we discussed in column buckling.

The value of C_b can be calculated from the following equation:

$$C_b = 1.75 + 1.05(M_1/M_2) + 0.3(M_1/M_2)^2 \le 2.3$$

where

M_1 = smaller end moment in unbraced segment

M_2 = larger end moment in unbraced segment

The ratio of M_1/M_2 is positive when the unbraced segment has reverse curvature, and the ratio is negative when the unbraced segment is bent in single curvature. C_b can be taken as 1.0 for segments in which a moment within the unbraced length is larger than at its ends and for cantilever beams. Therefore, in an unbraced beam the value of C_b will always be equal to 1.0.

The effect of the modifier C_b as it becomes greater than 1.0 is to account for the increased resistance to lateral torsional buckling by increasing the moment capacity.

The following example will demonstrate the use of the bending coefficient, C_b.

EXAMPLE 6.6

Calculate the adequacy of a W 12 × 96 that is 30 feet long and braced as shown in the figure. The beam carries a total load of 3 kips per foot and is made from A36 steel.

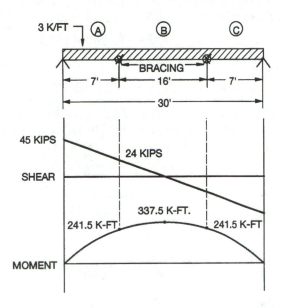

From the moment diagram we can see that segments A and C have $C_b = 1.75$, by calculating out the C_b formula (since $M_1 = 0$ and $M_2 = 241.5$ kip-ft). Segment 2 has $C_b = 1.0$ (since the moment at any point within that span is larger than it is at its ends). Therefore, segment 2, with an unbraced length, L_b, of 16 feet and a $C_b = 1.0$, is critical, since it is longer and has less resistance to lateral buckling due to the moment gradient.

Next, we must compare the critical unbraced length ($l_b = 16$ ft) to the smallest L_c value to determine from which category the allowable stress will come. Calculating L_c as follows:

$$L_c = 76 \times 12.16 / \sqrt{36} = 154 \text{ in. or } 12.8 \text{ ft}$$

$$L_c = 20,000 / (1.16)36 = 478 \text{ in. or } 39.9 \text{ ft}$$

Therefore, $L_c = 12.8$ ft, and since our unbraced length is greater than this value, the allowable stress value comes from category 3.

Since we are in category 3, calculate l/r_T as follows:

$$(16 \text{ ft} \times 12 \text{ in./ft}) \div 3.34 \text{ in.} = 57.5$$

With this l/r_T ratio we compare to the AISC dividing lines to see whether we use AISC Eq. (F1–6) or (F1–7). Remember that we always calculate AISC Eq. (F1–8). Comparing to the dividing lines we see:

$$53 \le l/r_T = 57.5 \le 119$$

Therefore, calculate the allowable stress from AISC Eqs. (F1–6) and (F1–8). From F1–6,

$$F_b = \{2/3 - [36(57.5)^2/(1{,}530{,}000 \times 1.0)]\}\ 36\ \text{ksi} = 21.2\ \text{ksi}$$

From F1–8,

$$F_b = 12{,}000 \times 1.0/[(16\ \text{ft} \times 12\ \text{in./ft})1.16] = 53.9\ \text{ksi}$$

The allowable stress is the larger—53.9 ksi—but it is larger than 0.60 F_y; therefore, use $F_b = 0.60\ F_y = 0.60\ (36\ \text{ksi}) = 21.6\ ksi$. Calculate the allowable moment that this beam can hold as follows:

$$M_{\text{allow}} = 21.6\ \text{ksi} \times 131\ \text{in.}^3 = 2829.6\ \text{kip-in.}\quad \text{or}\quad 235.8\ \text{kip-ft}$$

Therefore, the beam fails because 337.5 kip-ft > 235.8 kip-ft.

6.6   AISC Rolled-Beam Design Using ASD

The design of rolled-beam sections for flexural capacity follows the same logic as presented in the previous chapters. In designing beams, the problem is to select a section or, more correctly, to select a section modulus. The flexural formula can be rewritten as

$$S_{\text{req'd}} = \frac{M}{F_b}$$

The design of steel beams will hinge on which value to use for allowable stress. As was mentioned earlier, this depends on two factors: the section's compactness and the lateral support of its compression flange. Typically, the axis of bending will be the strong axis, and the condition of lateral support should be known. The unknowns are the compactness criteria (since they depend on a section not yet chosen) and compression flange buckling criteria ($l_b > L_c$ or $l_b \le L_c$). Remember that most rolled sections are typically compact.

This section will present a method of rolled-beam design that assumes that the designer would not have access to the many design charts available in the AISC manual. (The following section will present a rolled-beam design method using some typical design charts available in the AISC manual.) The design method proposed in this section is a logical trial-and-error approach that will converge on the final design in a relatively quick fashion. The underlying assumption in the following procedure is that the beam to be designed will, in fact, be compact. If this is the case, the problem is to determine whether the section falls in category 1 (F_b = 0.66 F_y) or category 3 (F_b = 0.60 F_y or lower). Remember that this determination is made by comparing the beam's unbraced length, l_b, to the smaller of the values of L_c. The values of L_c are based only on the flange width, b_f, the d/A_f term, and the strength of steel to be used. Therefore, by setting the beam's unbraced length, l_b, equal to the values for L_c we can ascertain the minimum flange width, $b_{f\,min}$, and the maximum d/A_f ratio, $d/A_{f\,max}$, that would allow our beam to be in category 1. By looking at the available section tables, we can then see what chance our beam has of being in category 1 and therefore our beginning our procedure using an allowable bending stress value of .66 F_y. Conversely, if our beam falls in category 3, the assumption of allowable bending stress should be 0.60 F_y or lower.

To reiterate, this procedure is a trial-and-error solution, but it will prove to be quick if we are familiar with the applicable specification requirements. The design steps for the selection of economical rolled-beam sections are outlined below.

1. Given maximum unbraced length, l_b, set this equal to the equations for L_c.
2. Solve for minimum flange width ($b_{f\,min}$) and maximum d/A_f ratio ($d/A_{f\,max}$), which would allow a section to fall in category 1.
3. Assuming a compact section, try to gauge the probability that a section would fall into category 1 or category 3. (Remember, if you believe the section to be chosen is a category 1 candidate, then assume F_b = 0.66 F_y. If you believe it to be a category 3 candidate, then assume F_b = 0.60 F_y or lower. Either of these assumptions must finally be proven.)
4. Calculate the section modulus required, $S_{x\,req'd}$, using the assumption of allowable bending stress from step 3:

$$S_{req'd} = \frac{M}{F_b}$$

5. Choose a section based on S_x, and check the criteria regarding b_f, d/A_f, and compactness. Calculate the section's allowable bending stress, F_b, and also calculate its moment capacity ($F_b \times S_x$) to make sure that it is greater than the applied moment (but not too much greater).

The following problems will illustrate the use of this technique.

EXAMPLE 6.7

Choose the most economical W12 section to hold a uniform load of 3 kips per foot over a 16-ft simple span. The steel is A36, and the beam is braced at midspan. Neglect self-weight.

Calculating the moment applied to the beam, we find the beam must hold a design moment as follows:

$$M_{design} = wl^2/8 = \frac{3\,kips/ft\,(16\,ft)^2}{8} = 96\,kip\text{-}ft$$

Setting the beam's unbraced length, l_b, equal to the L_c equations, we can solve for $b_{f\,min}$ and $d/A_{f\,max}$

$$l_b = 8\,ft \quad or \quad 96\,in. \qquad 96\,in. = 76 b_f/\sqrt{36} \qquad b_{f\,min} = 7.6\,in.$$

$$96\,in. = \frac{20000}{(d/A_f)36} \qquad d/A_{f\,max} = 5.79$$

Looking at the range of possibilities for W12 sections, we see that almost all W12 sections have a flange width over the minimum and d/A_f ratios lower than the maximum. In fact, any section over a W 12 × 35 will fall in category 1. Therefore, because there is a good probability that any beam selected will fall in category 1, assume $F_b = 0.66$ (36 ksi) = 23.76 ksi.

Calculate a required section modulus based on this assumption:

$$S_{required} = \frac{96\,kip\text{-}ft(12\,in./ft)}{23.76\,ksi} = 48.5\,in.^3$$

From AISC section tables, select a W 12 × 40 ($S_x = 51.9\,in.^3$) and check applicable unbraced length and compactness requirements to ensure that the assumed allowable stress of 0.66 F_y is correct.

$b_f = 8.005$ in. (actual) > 7.6 in. (minimum)   OK.

$d/A_f = 2.9$ (actual) < 5.78 (maximum)   OK.

$$\frac{b_f}{2t_f} = 7.8 < 65/\sqrt{F_y} \qquad \text{Flanges are compact — OK.}$$

$$\frac{d}{t_w} = 40.5 < 640/\sqrt{F_y} \qquad \text{Web is compact — OK.}$$

Calculate the allowable moment capacity to prove that a W 12 × 40 does supply the necessary resistance.

$$M_{allow} = 23.76\,ksi \times 51.9\,in.^3 = 1233.1\,kip\text{-}in. \quad or \quad 102.8\,kip\text{-}ft$$

Since 102.8 kip-ft > 96 kip-ft, a W 12 × 40 works and is economical.

The student should realize that self-weight cannot be neglected in real-world problems and that usually some estimation of this weight is made before calculating the design moment.

When the $b_{f\,min}$ and $d/A_{f\,max}$ ratios are such that a number of sections do not meet these criteria, we must realize that the chance of a section falling in category 1 is diminishing. Therefore, at some point we can assume that a design will fall in category 3. When assuming an allowable bending stress (F_b) in this category, the designer must realize that the maximum allowable stress is 0.60 F_y and it may even be much lower. The author's preference is to start with an allowable stress level of 0.60 F_y for the first trial and see how that assumption works in the problem at hand. If this assumption is too high, the author will adjust it to some lower level based on the result of the first trial.

EXAMPLE 6.8

Choose the most economical W24 section to carry 5 kips/ft over an unbraced span of 20 ft. The steel is A36; neglect self-weight.

$$M_{design} = wl^2/8 = \frac{(5 \text{ kips/ft})(20 \text{ ft})^2}{8} = 250 \text{ kip-ft}$$

$$l_b = 20 \text{ ft or } 240 \text{ in.} \quad 240 \text{ in.} = 76b_f / \sqrt{36} \quad b_{f\,min} = 18.95 \text{ in.}$$

and $d/A_f = 2.31$. (Checking the AISC section tables, we find no W24 sections with 19-in.-wide flanges, although most W24's have a d/A_f less than that required.)

Assuming the flange criteria will control, the probability of a section falling in category 1 is nil. Assume $F_b = 0.60 \, F_y$:

$$S_{required} = \frac{250 \text{ kip-ft} \times 12 \text{ in./ft}}{21.6 \text{ ksi}} = 139 \text{ in.}^3$$

From the AISC tables, a W 24 × 68 has a $S_x = 154$ in.³. Let's try this section. Calculate L_c:

$$L_c = \frac{76(8.97)}{\sqrt{36}} = 113.6 \text{ in. or } 9.5 \text{ ft} \quad \text{(This ends up controlling.)}$$

$$L_c = \frac{20000}{4.52(36)} = 122.9 \text{ in. or } 10.2 \text{ ft}$$

Since $l_b > L_c$, use AISC Eqs. (F1–6), (F1–7), and (F1–8) to calculate F_b:

$$\frac{l}{r_T} = \frac{20 \text{ ft} \times 12 \text{ in./ft}}{2.26} = 106.2 \quad C_b = 1.0 \quad \text{(simple, unbraced)}$$

Since $53 < 106.2 < 119$, use Eq. (F1–6):

$$F_b = \{2/3 - [36(106.2)^2/(1{,}530{,}000 \times 1.0)]\} \times 36 = 14.45 \text{ ksi}$$

Check Eq. (F1–8):

$$F_b = \frac{12{,}000(1)}{(20 \times 12)(4.52)} = 11.06 \text{ ksi}$$

Use the larger of the two, which is $F_b = 14.45$ ksi (which is usable since it is less than $0.60 \, F_y$).

Check the moment capacity:

$$M_{\text{allow}} = 154 \text{ in.}^3 \times 14.45 \text{ ksi} = 2225 \text{ kip-in.} \quad \text{or} \quad 185.5 \text{ kip-ft}$$

Since 185.5 kip-ft < 250 kip-ft, the W 24 × 68 is no good. Therefore, recalculate using an assumption lower than $0.60 \, F_y$ (say, 16 ksi):

$$S_{\text{required}} = \frac{250 \text{ kip-ft} \times 12 \text{ in./ft}}{16 \text{ ksi}} = 188 \text{ in.}^3$$

Choose W 24 × 84 ($S_x = 196$ in.^{3}) $l_b > L_c$ and from the AISC equations, $F_b = 14.86$ ksi (from a procedure similar to that already outlined with AISC Eq. (F1–6) controlling).

$$M_{\text{allow}} = 14.86 \text{ ksi} \times 196 \text{ in.}^3 = 2913 \text{ kip-in.} \quad \text{or} \quad 243 \text{ kip-ft, which is} <$$

250 kip-ft. Still no good, but just under. Therefore a W 24 × 94 will certainly work (go ahead and check just to be sure).

6.7   Preliminary Design Using the AISC Charts

As mentioned earlier, many charts and aids are available to a steel designer. One of the most commonly used charts, to help in the design of rolled-steel beams, is published in Part 2 of the AISC manual (pages 2–146 through 2–211). A sample of these charts is provided in Figure 6–8.

These charts have plotted the total allowable moment that wide-flange sections (and M-sections) can carry based on their unbraced length, l_b. The charts given in the AISC manual can be used for the aforementioned shapes made from steels with $F_y = 36$ ksi and $F_y = 50$ ksi. They are graphical representations utilizing the three categories and equations mentioned earlier in this chapter and are to be used when the bending coefficient, C_b, is equal to 1.0.

To use the AISC charts, a designer can simply enter the unbraced length, l_b, on the bottom scale and intersect that with the needed moment of resistance on the vertical scale. Any beam listed above and to the left of this intersection will satisfy the necessary requirements (considering that $C_b = 1.0$). Solid lines on the graph indicate the most economical section by weight in a given region, whereas dashed lines indicate that lighter sections will satisfy given strength requirements. The author would encourage readers to use this as a preliminary design step because the fundamentals of beam behavior must still be fully understood. Items such as compactness and the bending coefficient must still be checked to ensure proper

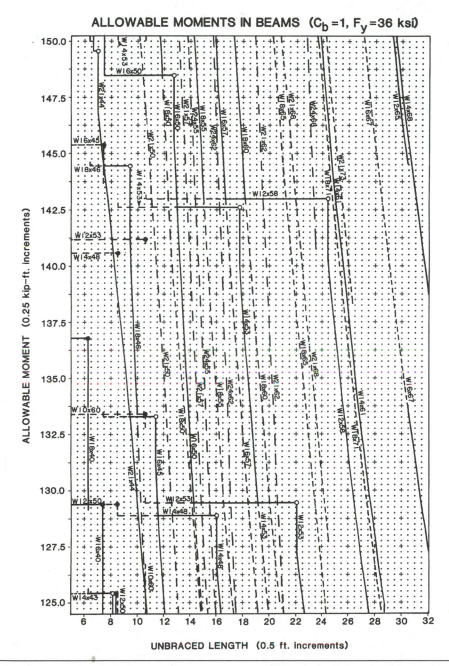

Figure 6–8   Beam Design Chart. (Courtesy of the American Institute of Steel Construction, Inc.)

usage of these charts. The open circles that appear on the line for an individual section indicate the length (sometimes referred to as L_u) over which a section's allowable stress falls below $0.60\ F_y$. The closed circles that may appear on the line for an individual section indicate that section's value of L_c.

The following example will illustrate the use of the AISC beam charts.

EXAMPLE 6.9

Design the lightest W-section to hold 5 kips per ft over a 15-ft unbraced simple span. The steel is A36, and the beam weight is to be neglected.

To begin, let's calculate the moment that needs to be resisted by the beam to be selected:

$$M = wl^2/8 = (5\ \text{kips/ft})(15\ \text{ft})^2/8 = 140.63\ \text{kip-ft}$$

Also, realize that we have steel with $F_y = 36$ ksi and $C_b = 1.0$, since the beam was simply supported and unbraced. Entering the charts with an unbraced length of 15 ft and a moment of 140.63 kip-ft (which happens to be the same one as shown in Figure 6–8), the first solid line that lies above this intersection is for a W 14 × 53. At this unbraced length, this section can hold approximately 142.5 kip-ft and the section is compact (although it does not matter since it falls in category 3).

6.8   AISC Requirements for Shear

Shear can be thought of as the sum of the loads on a beam up to the point under consideration. Therefore, shear forces will generally be the largest over the supports. From our discussion about beam behavior earlier in this chapter, we know that the maximum shear stress over a beam's cross section occurs at the neutral axis. Consequently, the focus of shear forces on rolled-steel sections concentrates on the beam's web and near the beam's supports.

Generally, the design of rolled-steel beams is not controlled by shear. Only when we have short, very heavily loaded beams, or when coping occurs, can shear criteria become a factor.

The AISC criteria for calculating shear stress over a rolled-beam cross section is given by the following equation:

$$f_v = \frac{V}{dt_w}$$

where

f_v = actual shear stress

V = shear at point under consideration (usually maximum)

d = beam depth

t_w = web thickness

This formula is based on an assumption of a uniform shear stress over the entire web. In reality, if we used the more "exact" shear stress formula ($f_v = VQ/Ib$), the results would typically show only a small variation from middepth to the junction of the flange and web. However, the exact formula usually gives somewhat higher results. Most designers typically ignore this discrepancy due to the fact that shear generally does not control the design.

The allowable stress for rolled sections, given in the AISC specifications (Section F4), is 0.40 F_y for beams that have a h/t_w ratio less than or equal to $380/\sqrt{F_y}$. (The h variable is the clear distance between flanges, but with rolled beams we can default to the conservative value of beam depth, d, since this will lead to a larger ratio.) The student will begin to realize that most rolled beams meet this criterion even for a high-strength steel.

If plate girders are used, the student should realize stiffeners might be needed and are referred to AISC Sections F4, F5, and F6 in the specifications.

6.9 Weak Axis Bending

Earlier in this chapter, when discussing the factors that affect the AISC's value of allowable bending stress, the two main factors given were compactness and lateral support of the compression flange. Actually, during this discussion we assumed that bending would occur about the shape's strong axis, but there is a possibility that bending could occur about a shape's weak axis (especially in beam-columns).

If bending occurs about the weak axis, which has a higher shape factor, the specification recognizes this behavior with a higher allowable stress. The code criteria for I-shaped members that are bent about their weak axis are found in AISC Section F2 and are outlined below.

- For compact sections bent about weak axis,

  $F_b = 0.75\ F_y$

- For noncompact sections bent about weak axis,

  $F_b = 0.60\ F_y$

- For sections not meeting flange compactness criteria,

$$F_b = F_y\left(1.075 - 0.005\left(\frac{b_f}{2t_f}\right)\sqrt{F_y}\right)$$

A rolled W-section bent about its strong axis has an approximate 12% increase in capacity above its yield moment. When bent about their weak axis, these same W-sections have on average a 50% increase in capacity before all fibers along the cross section have yielded. Although this increase in bending about the weak axis of a member seems lucrative, members are rarely bent about the weak axis, since other properties are so much lower than those about the strong axis.

6.10   Open-Web Steel Joists

Open-web steel joists are small parallel chord, truss-type beams whose members are often composed of small angles, round bars, or other steel shapes. The first open-web steel joist was manufactured as a Warren truss-type structure in 1923. Since then, open-web steel joists have become a very common structural steel beam member, used in both floor and roof construction of small to medium-sized buildings (Figure 6–9).

Figure 6–9   Open-Web Steel Joists in Steel Construction. (Courtesy of the Steel Joist Institute.)

Open-web joists have been standardized to fall in a number of "series," which all define certain purposes. The K-series of open-web steel joists are made for typical loads, spanning lengths up to approximately 60 feet. These joist are typically 8 in. to 30 in. in depth and have chords made from steel with an F_y of 50,000 psi. The J-series joists are similar to those of the K-series but are made from steels having an F_y of 36,000 psi, thereby reducing the span length they can handle under a given load. The J-series joists have essentially been replaced with the advent of higher-strength steel joists (namely, the K-series). Also available are LH- and DLH-series, which are made from steel with an F_y of 50,000 psi and are manufactured for longer span lengths. The DLH-series have joists that are 52 in. to 72 in. in depth and can span up to 144 feet. This makes them ideal for roof framing in warehouses, offices, and other mid-sized structures.

In designating open-web steel joists, the following format is typically used:

20 K 7

where

20 = the approximate depth of the joist, in.

K = the series designation

7 = chord size

The Steel Joist Institute (as well as practically all manufacturers of open-web steel joists) publish load tables[3] that tabulate the safe uniform load that certain joists can support at certain span lengths. Such safe loads are typically given so as to maintain a deflection of no more than 1/360 of the span. A standard type of load table is reproduced in Figure 6–10. Such tables are superb for selection of joists under uniform load only, but further analysis must be given if other load cases are to be considered.

6.11 Summary

Rolled beams are one of the basic building blocks for steel construction. Failure may initiate by many methods in steel beams, depending on the stability or buckling of the compression flange. This instability is referred to as lateral torsional buckling and is associated with the unbraced length, l_b, of the compression flange. If a beam is prevented (usually by lateral bracing) from this instability, it will typically be able to reach its plastic moment.

Design of steel beams can involve trial-and-error procedures and may be supplemented by using some of the many design aids that are now available. Shear failure on steel beams is typically not a controlling feature of steel design, unless the beams are very short, heavily loaded, or have coped sections. Open-web steel joists are another subset of steel beam design and are commonly used in light, industrial, and commercial projects.

STANDARD LOAD TABLE

Based on a Maximum Allowable Tensile Stress of 30,000 psi

OPEN WEB STEEL JOISTS, K-SERIES

Adopted by the Steel Joist Institute November 4, 1985; Revised to May 19, 1987.

The black figures in the following table give the TOTAL safe uniformly distributed load-carrying capacities, in pounds per linear foot, of K- Series Steel Joists. The weight of DEAD loads, including the joists, must be deducted to determine the LIVE load-carrying capacities of the joists. The load table may be used for parallel chord joists installed to a maximum slope of 1/2 inch per foot.

The figures shown in RED in this load table are the LIVE loads per linear foot of joist which will produce an approximate deflection of 1/360 of the span. LIVE loads which will produce a deflection of 1/240 of the span may be obtained by multiplying the figures in RED by 1.5. In no case shall the TOTAL load capacity of the joists be exceeded.

The approximate joist weights per linear foot shown in these tables do **not** include accessories.

The approximate moment of inertia of the joist, in inches[4] is: $I_J = 26.767 (W_{LL}) (L^3) (10^{-6})$, where W_{LL} = RED figure in the Load Table; L = (Span - 0.33), in feet.

For the proper handling of concentrated and/or varying loads, see Section 5.5 in the Recommended Code of Standard Practice (page 65).

Joist Designation	8K1	10K1	12K1	12K3	12K5	14K1	14K3	14K4	14K6	16K2	16K3	16K4	16K5	16K6	16K7	16K9
Depth (In.)	8	10	12	12	12	14	14	14	14	16	16	16	16	16	16	16
Approx. Wt. (lbs./ft.)	5.1	5.0	5.0	5.7	7.1	5.2	6.0	6.7	7.7	5.5	6.3	7.0	7.5	8.1	8.6	10.0
Span (ft.)																
8	550/550															
9	550/550															
10	550/480	550/550														
11	532/377	550/542														
12	444/288	550/455	550/550	550/550	550/550											
13	377/225	479/363	550/510	550/510	550/510											
14	324/179	412/289	500/425	550/463	550/463	550/550	550/550	550/550	550/550							
15	281/145	358/234	434/344	543/428	550/434	511/475	550/507	550/507	550/507							
16	246/119	313/192	380/282	476/351	550/396	448/390	550/467	550/467	550/467	550/550	550/550	550/550	550/550	550/550	550/550	550/550
17		277/159	336/234	420/291	550/366	395/324	495/404	550/443	550/443	512/488	550/526	550/526	550/526	550/526	550/526	550/526
18		246/134	299/197	374/245	507/317	352/272	441/339	530/397	550/408	456/409	508/456	550/490	550/490	550/490	550/490	550/490
19		221/113	268/167	335/207	454/269	315/230	395/287	475/336	550/383	408/347	455/386	547/452	550/455	550/455	550/455	550/455
20		199/97	241/142	302/177	409/230	284/197	356/246	428/287	525/347	368/297	410/330	493/386	550/426	550/426	550/426	550/426
21			218/123	273/153	370/198	257/170	322/212	388/248	475/299	333/255	371/285	447/333	503/373	548/405	550/406	550/406
22			199/106	249/132	337/172	234/147	293/184	353/215	432/259	303/222	337/247	406/289	458/323	498/351	550/385	550/385
23			181/93	227/116	308/150	214/128	268/160	322/188	395/226	277/194	308/216	371/252	418/282	455/307	507/339	550/363
24			166/81	208/101	282/132	196/113	245/141	295/165	362/199	254/170	283/189	340/221	384/248	418/269	465/298	550/346
25						180/100	226/124	272/145	334/175	234/150	260/167	313/195	353/219	384/238	428/263	514/311
26						166/88	209/109	251/129	308/156	216/133	240/148	289/173	326/194	355/211	395/233	474/276
27						154/79	193/98	233/115	285/139	200/119	223/132	268/155	302/173	329/188	366/208	439/246
28						143/70	180/88	216/103	265/124	186/106	207/118	249/138	281/155	306/168	340/186	408/220
29										173/95	193/106	232/124	261/139	285/151	317/167	380/198
30										161/86	180/96	216/112	244/126	266/137	296/151	355/178
31										151/78	168/87	203/101	228/114	249/124	277/137	332/161
32										142/71	158/79	190/92	214/103	233/112	259/124	311/147

Figure 6–10 Typical Open-Web Steel Joist Load Table. (Courtesy of the Steel Joist Institute.)

EXERCISES

1. A W 12 × 50 has to support a moment of 100 kip-ft. Calculate the section modulus and bending stress, if for the beam I_x = 394 in.^{4} and depth = 12.19 in.

2. Calculate the bending stress for a W 24 × 84 that has to carry a maximum moment of 200 kip-ft over a 16-ft span that is fully supported. The steel is A36.

3. Explain the differences between local buckling and lateral torsional buckling. Discuss the factors relating to each type of behavior.

4. Explain how the bending coefficient, C_b, affects the AISC allowable bending stress equations. Is a higher value of C_b good or bad as far as beam behavior is concerned?

5. A W 12 × 120 is used as a beam that is 20 feet long and braced along its compression flange at 5-ft intervals. If the maximum moment is 300 kip-ft (occurring at the beam's midpoint) and the steel is A36, will the beam be adequate per AISC bending stress criteria?

6. Using the information in Exercise 5, calculate the maximum uniform load that the beam can safely carry.

7. Check the adequacy of a W 24 × 84 to hold 350 kip-ft that is 20 feet long and unbraced along its compression flange. The beam is made from A36 steel and the load is assumed to be uniform.

8. Recheck the beam in Exercise 7, this time assuming the beam is braced at its midpoint.

9. Design the most economical W12 section to carry a maximum moment of 250 kip-ft located at midspan. The beam is 20 ft long and is braced at midspan; the steel is A36. Check shear and self-weight. The beam is assumed to be uniformly loaded.

10. For the beam shown in the accompanying figure, design the lightest W18 section, if it is
    a. Unbraced
    b. Braced at midspan
    c. Braced at quarter points
    Steel is A36.

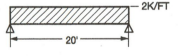

11. Design the steel support beam in the basement of the small ranch house outlined in the figure. The floor system consists of 2 × 10's @ 16 in. o.c. with 3/4-in. plywood. Floor live loads are 40 psf, and the steel is A36. Consider self-weight of all components, and consider the beam unbraced.

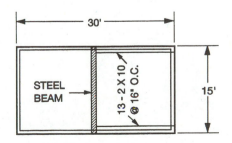

PLAN OF FOUNDATION

12. Using the load table as found in Figure 6–10, select an open-web steel joist to span 20 ft if the distributive area to that joist is 2 ft. The uniform dead load is 100 psf, and the uniform live load is 150 psf.

REFERENCES

1. Augustin Mrazih, *Plastic Design of Steel Structures* (London: Ellis Horwood Limited, 1987), pp. 94–95.
2. Robert Disque, *Applied Plastic Design in Steel* (New York: Van Nostrand Reinhold, 1971), p. 10.
3. "Standard Specifications, Load Tables and Weight Tables for Steel Joists and Joists Girders," Steel Joist Institute, Myrtle Beach, S.C., 1992.

CHAPTER

7

SERVICEABILITY AND DEFLECTION REQUIREMENTS

7.1   Introduction to Serviceability

The AISC specification defines serviceability as "a state in which the function of a building, its appearance, maintainability, durability, and comfort of its occupants are preserved under normal usage."[1] A building or member must function as its occupants wish, or it will fail to perform its function adequately. A serviceability failure can be as much of a problem as a strength-based failure, because the structure may not be used in the manner intended. Serviceability items, such as limiting the maximum beam deflection—rather than the standard strength criteria—may control the design of a member.

The topic of serviceability is given secondary status by some in the design profession, because it does not involve absolute quantities but rather deals with *quality* of acceptable behavior. While strength criteria are rather straightforward (actual stress ≤ allowable stress), serviceability criteria usually rely heavily on the designer's personal judgment (how much floor vibration is acceptable?). Further, a serviceability decision that works well on one particular project may not work well on another.

Common items that fall under the realm of serviceability include deflection, vibration, lateral drift, and overall appearance. The relationship of each item with serviceability will be discussed briefly.

Deflection

As members are loaded, they will deflect. If these deflections become excessive, they can lead to problems associated with appearance such as cracking or bulging in ceilings and walls. Other problems caused by excessive deflections may include improper functioning of other building components such as doors and windows. Excessive deflections may also lead to strength failures (as in the case of ponding, which will be discussed later in this chapter). Deflection is the most common serviceability-based design item and will be the focus of the next section.

Vibration

Members subjected to certain live loads (such as wind gusts, earthquakes, moving vehicles, and aerobic classes) will tend to vibrate under such loads. As long as the vibration does not become excessive, it is usually of little concern. However, when the frequency of the live load approaches that of the structure's natural frequency, vibration can reach unacceptable levels. One such case occurred at a restaurant/dance club in Flushing, New York, when floor vibrations caused by the dancing greatly alarmed the dining patrons.[2] This problem, although not constituting a strength-based structural defect, caused the building to be unacceptable for its intended functional use. Excessive vibration may also cause serviceability problems in industrial applications, where precisely tuned mechanical systems cannot tolerate movements over some limiting amount.

Lateral Drift

Lateral drift generally refers to the swaying of multistory buildings under wind loads. Large buildings, such as the World Trade Center in New York City, will sway laterally by as much as three feet in a storm with high winds. During such storms, lateral drifting must not be perceptible to the occupants of the building, to minimize any anxiety regarding the structure's integrity. Excessive drifting might also damage other building components such as doors and windows.

Overall Appearance

Excessive cracking in walls or ceilings, bulges or gaps in exterior facades, and damage due to improper drainage, all project a poor appearance to the casual observer. Such an appearance does nothing to instill confidence in the structural integrity of the building and therefore should be considered a serviceability aspect in any design.

7.2   Deflection Calculations and Requirements

The computation of beam deflection is necessary to ensure the serviceability of a structure, as was discussed in the previous section. Usually, in civil engineering structures, the deflections are known to be relatively small compared to the overall dimensions of the section and its span length. However, since a deflection criterion may control in some instances, the structural designer needs to be able to calculate the anticipated deflection in order to compare this value to the limiting values as set forth in the applicable building code.

As a beam is subjected to load, it will deflect. Deflection (δ) can simply be thought of as the result of the change of rotation or slope (θ) that occurs between different points along a beam. In fact, deflection can be defined as the *sum of incremental changes in rotation* along a beam. The rotation or slope change that a beam undergoes between two points is a function of the moment (at that point) and the beam's flexural stiffness (EI). This slope relationship is typically stated as follows:

$$\Delta\theta = \sum (M/EI)\, \Delta x$$

where

$\Delta\theta$ = incremental change of slope between two points

Δx = incremental x distance under consideration

Using this information, how do we calculate beam deflection? Actually, many methods are available to us, based on the relationships that exist between the bending moment on a beam of known rigidity and the corresponding rotation (slope) and deflection that this moment causes. Such methods of deflection calculation include the moment–area method, the conjugate beam method, and the double integration method. Possibly, the easiest of these techniques is the moment–area method. This method should be fully covered in any strength of materials text; therefore, only a limited explanation of this technique will follow.

The moment–area method for calculating deflection is based on the principle that the change of slope between any two points on a deflection curve is equal to the area between those two points on the M/EI diagram. For illustrative purposes, these points are A and B in Figure 7–1. (It should be noted that for a beam made of one material and constant cross section, the M/EI diagram is the same shape as the beam's moment diagram.) Since the total change of slope between points A and B is the basis for finding deflection, the shaded area between those points on the M/EI diagram must be found. This area for small increments of the M/EI diagram is approximately rectangular, and therefore the change in slope ($\Delta\theta$) is simply the value of M/EI multiplied by the incremental x distance (Δx).

The deflection (δ) of the beam at point A can be looked at as the elevation difference (tangent deviation) at various points relative to one another along the beam. This tangent deviation of one point relative to another (let's assume B relative to O) is found by intersecting a vertical line through B with a tangent line from O. Due to bending, the tangent deviation of the B relative to O is shown in Figure 7–1 and is calculated as follows:

$$t_{B/O} = x_B\,(\Delta\theta)$$

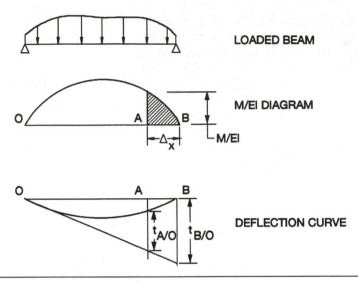

Figure 7–1 Relationships Involved in the Moment–Area Method.

where x_B = the centroidal distance from B to the area bounded by B–O under the M/EI diagram.

Substituting the equivalent formula for $\Delta\theta$, the formula may be written as follows:

$$t_{B/O} = \Sigma\ (M/EI\ \Delta x)\ x_B$$

The deflection at some point (such as A) may then be found by taking the tangent deviation from A to O ($t_{A/O}$) and subtracting that from the weighted value of tangent deviation from B to O ($t_{B/O}$).

The following example will demonstrate the use of this method in calculating deflection.

EXAMPLE 7.1

Using the moment–area method, derive the formula for deflection at the center of a simply supported beam with a concentrated load at its midpoint.

The first order of business is to construct the M/EI diagram for such a case, as shown in the figure. Notice the M/EI diagram is simply the moment diagram divided by the flexural stiffness.

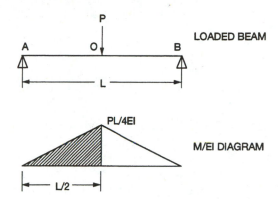

Since the deflection at the beam's center is desired, consider the amount of deflection to be equal to $1/2\ t_{B/A}$ minus $t_{O/A}$. Working with the two triangular areas under the moment diagram ($\Delta\theta$), list these areas (between these points B and A) as follows:

Area 1 $= (1/2) \times (L/2) \times (PL/4EI)$

$\qquad = PL^2/16EI$

Area 2 $=$ SAME

Now consider the distance from the centroid of that area to the point at which the tangent is taken (point B for $t_{B/A}$ and point O for $t_{O/A}$) In this case that distance would be as follows:

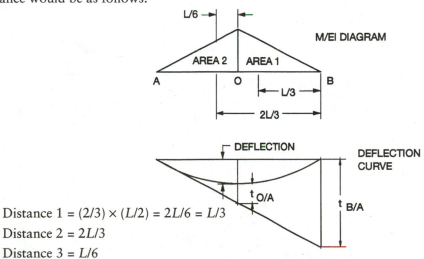

Distance 1 $= (2/3) \times (L/2) = 2L/6 = L/3$

Distance 2 $= 2L/3$

Distance 3 $= L/6$

Therefore, the deflection can be calculated as follows:

$\qquad \delta = (1/2)\ t_{B/A} - t_{O/A}$

where

$$t_{B/A} = (\text{Area 1} \times \text{Dist. 1}) + (\text{Area 2} \times \text{Dist. 2})$$
$$t_{O/A} = (\text{Area 2} \times \text{Dist. 3})$$

or

$$\delta = 3PL^3/96EI - PL^3/96EI = PL^3/48\ EI$$

Although this method can be used easily in many beam deflection cases, for typical beam loadings, there already exist deflection formulas (derived from these aforementioned methods). Such formulas are found in many references such as the AISC *Manual of Steel Construction*. A partial list of beam deflection formulas for some of the more common beam loadings is given in Figure 7–2. (For a more comprehensive listing of deflection formulas, please refer to pages 2–296 through 2–307 of the AISC manual.)

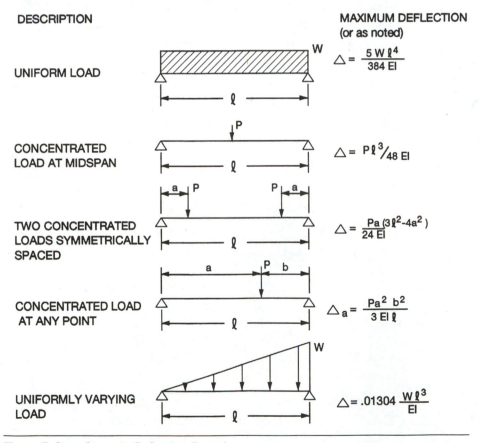

DESCRIPTION

MAXIMUM DEFLECTION
(or as noted)

UNIFORM LOAD
$$\triangle = \frac{5\,W\,\ell^4}{384\,EI}$$

CONCENTRATED LOAD AT MIDSPAN
$$\triangle = P\ell^3/48\,EI$$

TWO CONCENTRATED LOADS SYMMETRICALLY SPACED
$$\triangle = \frac{Pa\,(3\ell^2 - 4a^2)}{24\,EI}$$

CONCENTRATED LOAD AT ANY POINT
$$\triangle_a = \frac{Pa^2\,b^2}{3\,EI\,\ell}$$

UNIFORMLY VARYING LOAD
$$\triangle = .01304\,\frac{W\ell^3}{EI}$$

Figure 7–2   Common Deflection Formulas.

The following examples will illustrate how some of these deflection formulas are used. The most important reminder when using these formulas: *Keep consistent units!* Remember that deflection is usually measured in inches. Therefore, make sure all uniform loads are in units of pounds per inch or kips per inch (depending on what units the modulus of elasticity uses). All span lengths are also typically listed in units of inches.

EXAMPLE 7.2

Calculate the live load deflection of the 20-ft-long W 12 × 50 under a uniform live load of 3 kips per foot over its entire length.

In this problem, the deflection is calculated from the formula found in Figure 7–2. The variables needed for this calculation are as follows:

w = 3 kips/ft = 3 kips/ft ÷ (12 in./ft) or *.25 kips/in.*

l = 20 ft = *240 in.*

E = 29,000 ksi (since load is in units of kips)

I = 394 in.^{4} (W 12 × 50 about its strong axis)

Calculating deflection:

$\delta = 5wl^4/384EI$

δ = 5 × (.25 kips/in.) × (240 in.)^4/384(29,000 ksi)(394 in.^{4})

   = *.945 in.*

(*Note:* To include dead load deflection caused by the beam's self-weight, one would simply consider an additional 50 lb/ft of loading (or 4.17 lb/in.)

EXAMPLE 7.3

Calculate the live load deflection on the same beam as in Example 7.2 under a concentrated live load of 60 kips located at midspan.

Use the appropriate formula from Figure 7–2. The variables needed to calculate the formula are as follows:

l = 20 ft = 240 in.

P = 60 kips

I = 394 in.^{4}

E = 29,000 ksi

Using the formula:

$\delta = Pl^3/48\ EI$

$\delta = (60\ \text{kips})(240\ \text{in.})^3/48(29{,}000\ \text{ksi})(394\ \text{in.}^4) = 1.51\ \text{in.}$

Now that we have examined one way to calculate deflections, what limiting values of deflections are used in standard practice?

Usually, building codes and specifications will limit deflections to a fraction of a beam's span length. Most building codes, as well as the AISC specification in Chapter L, limit the live load deflection on beams under service load conditions to 1/360 of the span length (or $l/360$) in plastered construction. When dealing with beams in unplastered construction, this requirement is relaxed to 1/180 of the span length (or $l/180$) in roof assemblies and to 1/240 of the span length in floor assemblies. In comparison, the AASHTO specification[3] limits the live load deflection on steel bridge girders to 1/800 of the span length (or $l/800$) to maintain clearance height under typically heavy live loads.

The student should remember that the aforementioned deflection limits are recommended maximums that are meant to provide guidance (not a guarantee) in achieving a serviceable design. Designers must always use their personal judgment, taking into account the goals of a specific project. These values may be adjusted, either upward or downward, should that individual project warrant such a decision.

7.3 Deflection Strength Failure: Ponding

Up to now, we have discussed beam deflection solely as a serviceability concern. Usually, this is indeed the case. There is, however, one notable exception where deflection problems can actually be the catalyst behind a destructive strength-based failure. This one exception is a phenomenon referred to as **ponding**, the results of which can be devastating.

Ponding refers to the retention of water on a flat or semi-flat roof during periods of heavy rainfall. As the water accumulates on the roof faster than it can drain off, the roof deflects under the weight of the water, forming a bowl-shaped profile that enables the roof to retain even more water (Figure 7–3). The roof will keep deflecting, thereby holding more water, until it finally collapses.

Ponding has become more of a problem with the advent of flexible or "light" roof framing systems. More flexible roof beams and girders increase the chance that ponding may occur in a severe rainstorm. Therefore, the AISC requires that ponding criteria (found in Section K2 of the AISC specification) be met to ensure adequate stiffness for primary and secondary roof members. These criteria are beyond the scope of this introductory text; students who intend to continue the study of structural steel should investigate the specifications regarding this behavior.

Some lessons learned from ponding failures are more subtle in nature and

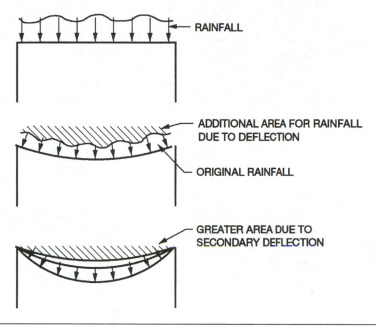

Figure 7–3   Progression of Ponding Failure.

more practically oriented than the aforementioned stiffness specifications. When a 12,000-square-foot roof section of an Internal Revenue Service warehouse collapsed in Philadelphia, one of the seemingly simple lessons to be learned was to locate roof drains away from column lines. This is because the columns do not deflect nearly as much as roof framing does and the areas adjacent to columns therefore remain "high" in regard to roof profile. Ponding will take place between columns and, therefore, the optimum location for roof drains would be at midspan between columns (if feasible).[4]

7.4   Man-Induced Vibration as a Serviceability Concern

With society placing an increasingly high value on entertainment and fitness in the last 20 years, vibration problems caused by human movement in structures are no longer a rare occurrence. Numerous gymnasiums, sporting arenas, and concert halls have experienced excessive vibration due to the rhythmic motions caused by either the participants or the spectators.[5] Aerobic classes, rock concerts, and periodic motions (such as the "wave") all impart a vibrational behavior to a structure. Although these vibrations are primarily a serviceability concern, strength aspects such as fatigue stresses must also be considered.

Excessive vibration can rattle windows and doors, generally making a person uncomfortable in occupying a structure. As the human motion (jumping, running, clapping of hands) becomes rhythmic, it begins to occur at a certain sustained frequency. These motions set the structure into vibration at its own particular frequency, called its natural frequency. If the periodic human motion has a frequency close to the structure's natural frequency, vibrational behavior tends to be amplified. This amplification is the behavior that greatly disturbs a casual observer.

Generally, there are a couple of different ways to alleviate vibrational problems should they occur in existing structures. The first is to stiffen the existing structure by adding intermediate beams, columns, or other bracing members. This stiffening tends to increase the structure's natural frequency, thereby avoiding the aforementioned amplification behavior. Another technique being used successfully in treating vibrational problems is the use of **tuned-mass dampers** (TMDs). Tuned-mass dampers are mechanical vibration control systems that work similarly to the shock absorber of an automobile. The system typically involves fastening a counterweight to the vibrating member using a spring and damper. As the member reaches its critical frequency, it produces a corresponding movement in the TMD (Figure 7–4). This movement in the TMD produces a "damping effect" on the vibrating member, reducing the vibrational behavior. This would be analogous to a person pushing on a swing before it reached the apex of its arc.

Other damping systems are presently being researched and show signs of being effective in controlling excessive vibrations. One notable method uses viscoelastic dampers, in which a viscous material converts mechanical energy, induced from the vibration, into heat.

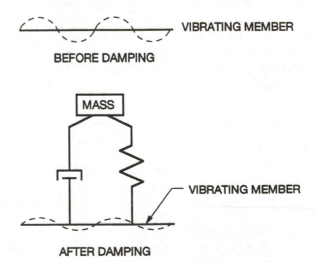

Figure 7–4   Schematic Illustration of a Tuned Mass Damper.

7.5   Summary

Serviceability can be thought of as the functioning of a structure in the manner it was designed. Some of the more typical problems of serviceability stem from excessive deflection, vibration, or lateral drift. In the worst-case scenario, serviceability problems may lead to catastrophic failure, such as a structure's collapse due to ponding.

The designer must be aware of the serviceability aspects of the particular structure and adjust specification requirements accordingly.

EXERCISES

1. Explain why serviceability should be very important for a structural designer. Why, then, is it sometimes overlooked?
2. Calculate the live load deflection caused by two 50-kip live loads, each located 5 ft in from their respective supports on a 25-ft simply supported beam. The beam is a W 12 × 120, and the self-weight is neglected. Would this deflection be acceptable under service floor load conditions using plastered construction?
3. Calculate the total deflection for the beam in Exercise 2, if dead load from the beam's self-weight is to be included. (*Hint*: superimpose the deflections caused by each load.)
4. Calculate the uniform live load that would cause a live load deflection of 1 in. in a W 12 × 106 that is 15 ft long. Recalculate for a beam that is 20 ft long.
5. Calculate the concentrated live load to cause a .50-in. live load deflection in a W 12 × 14 that is 22 ft long. This beam is commonly used in some areas for residential framing. Is there anything in your house that may cause such a deflection?
6. Refer back to Example 6.7 in Chapter 6. Calculate the live load deflection for the beam designed. Would this meet the plastered construction deflection criteria?
7. Design a W12 section to hold a uniform live load of 4.5 kips per foot over a simple span of 17 feet. The beam is braced at midspan and is made from A36 steel. Consider live load deflection criteria for unplastered floor load construction.
8. Design a W12 section to hold a concentrated live load of 30 kips at the center of a 20-ft-long, simply supported beam. The beam is to be made from A36 steel and to be unbraced except for the ends. In addition, the live load deflection must be limited to 1/1000 of the span.
9. Using the moment–area method, derive the deflection formula at the 1/4 point for a beam with a concentrated load at its center.

REFERENCES

1. *Manual of Steel Construction,* 9th ed. (Chicago: American Institute of Steel Construction, 1989).
2. Anthony C. Webster and Matthys P. Levy, "A Case of the Shakes," *Civil Engineering,* February 1992, p. 58.
3. *Standard Specifications for Highway Bridges,* 14th ed. (Washington, D.C.: American Association of State Highway and Transportation Officials, 1989).
4. Dov Kaminetsky, "Design and Construction Failures, Lessons Learned from Forensic Investigations" (New York: McGraw-Hill Publishing, 1991), p. 249.
5. H. Bachmann, "Case Studies of Structures with Man-Induced Vibrations," *Journal of Structural Engineering,* ASCE, vol. 118, no. 3, March 1992.

C H A P T E R

8

COMBINED BENDING
AND AXIAL MEMBERS:
BEAM-COLUMNS

8.1   Introduction

Up to this point, we have discussed structural steel members that are subjected to either direct axial stress (tension and compression members) or flexural stress (beams). We have considered the aforementioned stresses to act on these members to the exclusion of all others. But, in reality, will members be subjected to just one type of stress?

The simple answer is no. Structural members encounter many different combinations of stress under normal service conditions. Many times, to simplify the analysis and calculations, the student has probably assumed that such combinations are nonexistent. The most common combination of stresses occurs when a member is subjected simultaneously to axial and bending stresses. This situation occurs frequently in cases such as eccentrically loaded axial members, beams and columns subjected to lateral forces, or even truss members bending due to their own self-weight.

The most common case of combined axial and bending stress (and probably the most important) is that of an axially loaded compression member subjected to appreciable amounts of bending moment. Such members are referred to as **beam-columns.** Beam-columns are a serious concern because the moment applied to these members will cause a deflection, thereby creating more moment due to the effect of the axial load and the lateral displacement. This additional moment is referred to as secondary moment and can lead to further lateral displacement, which leads to more moment, and so on. Remembering the tendency of a column to laterally buckle under the application of compressive stress, we can readily accept the concern over beam-columns.

Members subjected to combined axial compression and bending stresses will be the focus of this chapter. The concern over members subjected to axial tension and bending is not as great, because the tension stress tends to "straighten out" the member, limiting lateral deflections and thereby reducing any secondary moment effect. Therefore, the following sections primarily will attempt to enlighten the student to only the fundamentals of beam-column behavior and design.

8.2   Potential Modes of Failure and Beam-Column Behavior

Since a beam-column is actually a mixture of two separate types of behaviors, it only stands to reason that the potential failure mechanisms may also reflect the aforementioned behaviors of these member types. As the axial load becomes ever-increasingly large on an individual beam-column, the failure mode will gravitate toward that of a true column. Likewise, as the bending moment grows increasingly larger, the failure mode will begin to resemble closely that of a true beam. The failure mechanisms of both a beam and column have been discussed at some length (and will not be fully reiterated here); the student is urged to review these mechanisms as needed. Because the most common failure mechanism in both beams and columns revolves around buckling of the compression elements, the lateral stability of the beam-column is the primary concern in the design and analysis of this member.

Behavior of the beam-column heightens this concern over lateral stability due to the fact that as bending moment is applied, the member will deflect laterally. This lateral displacement actually causes additional moment (called secondary moment) due to the eccentricity of the load with the column's longitudinal axis (Figure 8–1).

As this secondary moment appears, the beam-column is further deflected and the secondary moment increases until it reaches equilibrium. To account for this increase in moment due to the beam-column behavior, the AISC specification introduces an amplification factor. This amplification factor is as follows:

$$1/[1 - (f_a/F'_e)]$$

where

f_a = actual axial stress

F'_e = Euler buckling stress formula divided by a safety factor of 23/12

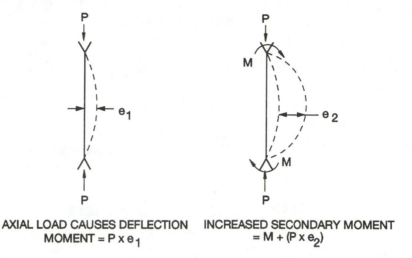

AXIAL LOAD CAUSES DEFLECTION
MOMENT = P x e_1

INCREASED SECONDARY MOMENT
= M + (P x e_2)

Figure 8-1   The Development of Secondary Moment.

This amplification factor will be larger than 1.0 and will be multiplied by the actual bending stress to effectively show an increase in the magnitude of applied moment. The AISC specification incorporates this amplification factor into its formulas, as further discussed in the next section.

In some instances, the aforementioned amplification factor overestimates the effect of the secondary moment. To counteract this possible overestimation, the AISC also introduces a modification factor, C_m. This modification factor will be 1.0 or less, depending on items such as presence of transverse loading and member end-restraint conditions. The general values of C_m are broken down into the following three categories:

1. C_m = .85; columns that are part of unbraced frames and are therefore subject to joint translation or sidesway.

2. C_m = .6 − .4(M_1/M_2); columns that are braced against sidesway and are not subjected to transverse loadings between their ends. The ratio M_1/M_2 is the ratio of smaller end moment to larger end moment. This ratio is negative if bent in single curvature and positive if bent in reverse curvature.

3. For columns that are braced against sidesway, C_m = 1.0 for members subjected to transverse loadings between unrestrained ends. C_m = .85 for members subjected to transverse loadings between restrained ends.

Although there are more refined methods to obtain modification factors for members subjected to transverse loadings (as discussed in the ASD Commentary), the preceding expressions are considered suitable for most applications.

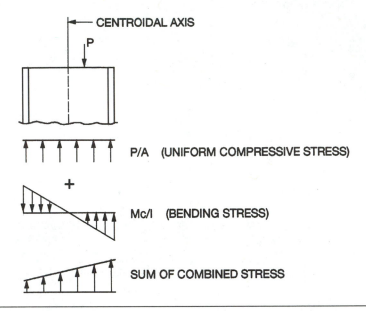

Figure 8–2   The Combined Stress Principle in an Eccentically Loaded Column.

8.3   Development of the Interaction Formulas

Having taken a course on strength of materials, the student no doubt remembers the subject of combined stresses. Under the topic of combined stresses, the principle of superpositioning was advanced, and the student learned that stresses of the same type can be added and subtracted numerically (see Figure 8–2).

When discussing combined bending and axial stresses, this principle led to the derivation of the combined stress formula, shown as follows:

$f = P/A \pm Mc/I$

where

P/A = axial stress

Mc/I = bending stress

Although approximate in nature (because it does not include the effects of secondary moments), this formula is widely accepted for the calculation of combined bending and axial stresses. Since we can roughly calculate actual combined stresses, how might combined allowable stress be handled in the realm of design?

The answer is that most design specifications disregard the idea of a combined allowable stress in favor of an expression known as an interaction formula. Most interaction formulas calculate ratios or percentages of actual stress to allowable stress for individual stress types and then add these ratios. If the ratio is less than 1.0, the combined stress is considered to be acceptable. This general interaction formula can be seen as follows:

$$f_a/F_a + f_b/F_b \leq 1$$

where

f_a, F_a = actual and allowable axial stress, respectively

f_b, F_b = actual and allowable bending stress, respectively

A good way to view this interaction formula is to consider the percent of allowable stress used in any given type of stress. For instance, if the actual axial stress was only 20% of its allowable value, then it would follow that the safe amount of actual bending stress would be 80% of the allowable bending value.

8.4   AISC Specifications for Beam-Columns

Chapter H in the AISC specification discusses the handling of members subjected to combined axial and bending stresses. Three interaction formulas regarding this behavior are presented in this chapter for use with beam-columns. When appreciable axial stress is applied to a beam-column (when the ratio of $f_a/F_a > .15$), the AISC specification checks the first two interaction formulas given. The first formula (H1–1) is a check of stability near the midpoint of the beam-column and takes into account the aforementioned moment amplification due to secondary effects. The second equation to be checked (H1–2) is used to satisfy stress requirements near the ends of the beam-column. The following equations also display terms for bending about both the strong and weak axes, although if bending occurs about one axis only, the other bending term would be eliminated:

$$\frac{f_a}{F_a} + \frac{C_{mx}f_{bx}}{\left(1-\dfrac{f_a}{F'_{ex}}\right)F_{bx}} + \frac{C_{my}f_{by}}{\left(1-\dfrac{f_a}{F'_{ey}}\right)F_{by}} \qquad \text{(AISC Eq. H1–1)}$$

$$\frac{f_a}{0.60F_y} + \frac{f_{bx}}{F_{bx}} + \frac{f_{by}}{F_{by}} \qquad \text{(AISC Eq. H1–2)}$$

If the axial stress on a beam-column is relatively small (when the ratio of $f_a/F_a \leq .15$), the AISC specification checks only the third equation (H1–3), which is shown next. In this case, the effects of secondary moment are insignificant; therefore, the moment amplification factor is disregarded.

$$\frac{f_a}{F_a} + \frac{f_{bx}}{F_{bx}} + \frac{f_{by}}{F_{by}} \qquad \text{(AISC Eq. H1–3)}$$

In these three interaction equations, the expressions for actual axial and bending stress (f_a, f_b), allowable axial and bending stress (F_a, F_b), and modification factor (C_m) are the same values mentioned previously in this chapter and throughout the book. The value F'_e is Euler's critical buckling stress divided by a safety factor of 23/12, which is taken in the plane of bending. The formula for F'_e follows, and the expressions for K (effective length factor), l_b (unbraced length), and r_b (radius of gyration) are equivalent to those discussed previously in this book.

$$F'_e = \frac{12\pi^2 E}{23\left(Kl_b / r_b\right)^2}$$

The following examples will demonstrate the use of these interaction formulas and clarify the evaluation procedure for beam-columns.

EXAMPLE 8.1

A W 12 × 120 is used to support the loads and moments, as shown in the figure, and is subjected to sidesway. Determine whether the member is adequate if the bending occurs about the weak axis. The column is assumed to be perfectly pinned ($K = 1.0$) in both the strong and weak directions, and no bracing is supplied. The steel is A36.

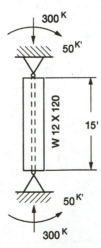

Initially, we should calculate the slenderness ratios in both the weak and strong directions (even though when unbraced in both directions, the weak axis will of course be critical).

$Kl_x/r_x = 1.0(15 \text{ ft} \times 12 \text{ in./ft})/5.51 \text{ in.} = 32.67$

$Kl_y/r_y = 1.0(15 \text{ ft} \times 12 \text{ in./ft})/3.13 \text{ in.} = 57.50$ (controls)

Calculating the allowable axial stress, F_a, from AISC Eq. (E2–1) (because $Kl_y/r_y < C_c$), we find that

F_a = 17.67 ksi

Calculating actual axial stress, f_a, we find that

$f_a = P/A$ = 300 kips/35.3 in.^{2} = 8.50 ksi

The ratio of f_a/F_a = 8.50/17.67 = .48, which is greater than .15. Therefore, use AISC Eqs. (H1–1) and (H1–2) to check beam-column adequacy.
    Calculate actual and allowable bending stresses about the weak axis:

$f_{by} = M/S_y$ = 50 kip-ft (12 in./ft)/56 in.^{3} = 10.71 ksi

$F_{by} = .75\ F_y$ (since compact) = .75(36 ksi) = 27 ksi

Calculate Euler buckling stress and the amplification factor:

F'_{ey} = 12(π^2)(29,000 ksi)/{23 × [1.0(15 ft × 12 in./ft)/3.13 in.]^2} = 45.15 ksi

Amp. factor = $1 - (f_a/F'_{ey})$ = 1 – (8.50 ksi/44.57 ksi) = .809
    (This amplifies the moment by being in the denominator of the
    AISC equations.)

$C_m = .85$ (column subjected to sidesway)

Now calculate AISC Eqs. (H1–1) and (H1–2):

$$\frac{f_a}{F_a} + \frac{C_{mx}f_{bx}}{\left(1-\dfrac{f_a}{F'_{ex}}\right)F_{bx}} + \frac{C_{my}f_{by}}{\left(1-\dfrac{f_a}{F'_{ey}}\right)F_{by}} \qquad \text{(AISC Eq. H1-1)}$$

(8.50 ksi/17.67 ksi) + 0 + (.85)(10.71 ksi)/[(.809)(27)] = .897 < 1.0   OK

$$\frac{f_a}{0.60\,F_y} + \frac{f_{bx}}{F_{bx}} + \frac{f_{by}}{F_{by}} \qquad \text{(AISC Eq. H1-2)}$$

(8.50 ksi/21.60 ksi) + 0 + (10.71 ksi/27 ksi) = .79 < 1.0   OK

Since the beam-column satisfies both equations, it is adequate under the applied loads.

EXAMPLE 8.2

The W 14 × 90 is subjected to the loads and moments shown in the figure and is prevented from sidesway. If bending takes place about the strong axis, determine the adequacy using the AISC interaction formulas. The ends are assumed to have $K = 1.0$, and the steel is A36.

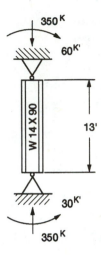

Initially calculate the slenderness ratios about each axis, although it is evident the weak axis will control given that no bracing is present.

Kl/r_x = 1.0(13 ft × 12 in./ft)/6.14 in. = 25.41

Kl/r_y = 1.0(13 ft × 12 in./ft)/3.70 in. = 42.16   (controls)

Now calculate the actual axial stress and the allowable axial stress based on the controlling slenderness ratio (using AISC Eq. E2–1):

$f_a = P/A$ = 350 kips/26.5 in.^{2} = 13.21 ksi

F_a = 19.01 ksi   (from AISC Eq. E2–1)

Since f_a/F_a = 13.21 ksi/19.01 ksi = .695, which is greater than .15, we have to again check the AISC interaction equations (H1–1) and (H1–2).

Now calculate the actual and allowable bending stresses about the strong axis:

$f_b = M/S_x$ = 60 kip-ft (12 in./ft)/143 in.^{3} = 5.03 ksi

$F_b = .66 F_y$ = .66(36 ksi) = 23.76 ksi   (the section is compact and $l_b < L_c$—category 1)

Calculate Euler's buckling stress about the strong axis, the amplification factor, and modification factor.

F'_{ex} = 12(π^2)(29,000 ksi)/[23(25.41)^2] = 231.3 ksi

Amp. factor = 1 – (13.21 ksi/231.3 ksi) = .943

C_m = .6 – .4(M_1/M_2) = .6 – .4(–30 kip-ft/60 kip-ft) = 0.8

(Since the end moment results in single curvature, the member is more likely to buckle. Therefore, the sign is negative. The sign is positive if the end moments create reverse curvature.)

Now calculate AISC Eq. H1–1 and H1–2:

$$\frac{f_a}{F_a} + \frac{C_{mx}f_{bx}}{\left(1-\dfrac{f_a}{F'_{ex}}\right)F_{bx}} + \frac{C_{my}f_{by}}{\left(1-\dfrac{f_a}{F'_{ey}}\right)F_{by}} \qquad \text{(AISC Eq. H1–1)}$$

(13.21 ksi/19.01 ksi) + (5.03 ksi)/[(23.76 ksi)] + 0 = .91 > 1.0   OK

$$\frac{f_a}{0.60\,F_y} + \frac{f_{bx}}{F_{bx}} + \frac{f_{by}}{F_{by}} \qquad \text{(AISC Eq. H1–2)}$$

(13.21 ksi/21.6 ksi)+ (5.03 ksi/23.76 ksi) + 0 = .82 < 1.0   OK

This member is adequate.

8.5   Design of Beam-Columns Using ASD

The design of beam-columns is truly a trial-and-error procedure. If the designer can select a good trial section at the very beginning, the work in obtaining a safe and economical member is greatly simplified. The most popular method of selecting a beam-column is the **equivalent axial load method.** This method replaces the axial load and moment that are applied to the column with a concentric axial load. The concentric axial load will be larger than the actual axial load but will produce approximately the same maximum stress.

This method uses equations to convert the applied bending moment into an estimated axial load, P'. The actual applied axial load, P_o, is then added to P' resulting in a fictitious axial load called the equivalent axial load, P_{eff}. This idea is

$$P_o + P' = P_{eff}$$

where

P_o = actual axial load

P' = estimated axial load caused by moment

P_{eff} = equivalent axial load

Although the AISC has equations that would estimate the equivalent axial load based on the assumptions used by the three interaction formulas (H1–1, H1–2, H1–3) separately, it is much faster to use the following combined approximation developed from these formulas:

$$P_{eff} = P_o + M_x m + M_y m U$$

Table 8–1   Values of m^* for Beam-Column Design ($F_y = 36$ ksi; $F_y = 50$ ksi)

F_y			36 ksi							50 ksi				
KL (ft)	10	12	14	16	18	20	22 & over	10	12	14	16	18	20	22 & over
						1st Approximation								
All shapes	2.4	2.3	2.2	2.2	2.1	2.0	1.9	2.4	2.3	2.2	2.0	1.9	1.8	1.7
						Subsequent Approximations								
W, S 4	3.6	2.6	1.9	1.6	—	—	—	2.7	1.9	1.6	1.6	—	—	—
W, S 5	3.9	3.2	2.4	1.9	1.5	1.4	—	3.3	2.4	1.8	1.6	1.4	1.4	—
W, S 6	3.2	2.7	2.3	2.0	1.9	1.6	1.5	3.0	2.5	2.2	1.9	1.8	1.6	1.5
W8	3.0	2.9	2.8	2.6	2.3	2.0	2.0	3.0	2.8	2.5	2.2	1.9	1.6	1.6
W10	2.6	2.5	2.5	2.4	2.3	2.1	2.0	2.5	2.5	2.4	2.3	2.1	1.9	1.7
W12	2.1	2.1	2.0	2.0	2.0	2.0	2.0	2.0	2.0	2.0	1.9	1.9	1.8	1.7
W14	1.8	1.7	1.7	1.7	1.7	1.7	1.7	1.8	1.7	1.7	1.7	1.7	1.7	1.7

Source: Courtesy of the American Institute of Steel Construction, Inc.
*Values of m are for $C_m = 0.85$. When C_m is other than 0.85, multiply the tabular value of m by $C_m/0.85$.

In this formula, P_o is the actual load expressed in kips and M_x and M_y are the bending moments expressed in kip-feet. If bending should occur about one axis only, the corresponding term regarding the other axis will then drop out. The value of m is taken from Table 8–1, and the value of U is taken from the column load tables. (The column load tables can be found in the AISC manual. The load tables for the W12 sections have been reprinted and are found in Appendix C of this book.) The procedure using this approximate formula is an iterative process that begins by taking the value of m from the chart in Table 8–1 and assuming the value of U to be equal to 3.0. After a section is selected, the section is checked against the applicable AISC interactions formulas (H1–1, H1–2, H1–3). With successive trials the values of m and U can be refined to a point where their values begin to stabilize.

The following example will utilize this procedure in the design of a beam-column.

EXAMPLE 8.3

Design the most economical W12 section subjected to the loads and moments shown in the figure. $K = 1.0$ in both strong and weak directions, and the steel is A36. Assume this member to be part of a frame that is subject to sidesway.

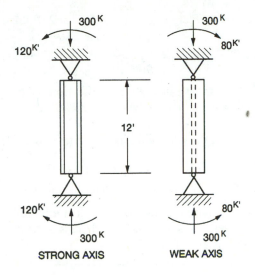

STRONG AXIS   WEAK AXIS

Using the equivalent load formula and choosing the first trial values of $m = 2.3$ (Table 8–1) and $U = 3.0$ (recommended), the equivalent load is as follows:

P_{eff} = 300 kips + 120 kip-ft (2.3) + 80 kip-ft (2.3)(3.0) = 1128 kips

From the column load tables for $Kl = 12$ ft, try a W 12×210. Checking this section with the interaction formulas, we find the following:

$f_a = P/A$ = 300 kips/61.8 in.^{2} = 4.85 ksi

Using $Kl/r_y = 1.0(12$ ft $\times 12$ in./ft)/3.28 in. = 43.90, we find that $F_a = 18.87$ ksi. (Since $f_a/F_a = .257$, check Eqs. H1–1 and H1–2.)

Calculate bending stresses about both axes:

$f_{bx} = M/S_x$ = 120 kip-ft (12 in./ft)/292 in.^{3} = 4.91 ksi
$f_{by} = M/S_y$ = 80 kip-ft (12 in./ft)/104 in.^{3} = 9.23 ksi

Find the allowable bending stresses:

The unbraced length, $l_b = 12$ ft and $L_c = 13.5$ ft
$F_{bx} = .66 \, F_y$ = 23.76 ksi   (since $l_b < L_c$ and its compact)
$F_{by} = .75 \, F_y$ = 27 ksi   (since weak axis bending)

Calculate amplification factors:

$F'_{ex} = 12\pi^2(29{,}000$ ksi$)/[23(24.45)^2]$ = 249.8 ksi
$F'_{ey} = 12\pi^2(29{,}000$ ksi$)/[23(43.9)^2]$ = 77.48 ksi

$$C_{mx}/1 - (f_a/F'_{ex}) = .85/[1 - (4.85 \text{ ksi}/249.8 \text{ ksi})]$$
$$= .867 \quad \text{Use } 1.0.$$
$$C_{my}/1 - (f_a/F'_{ey}) = .85/[1 - (4.85 \text{ ksi}/77.48 \text{ ksi})]$$
$$= .907 \quad \text{Use } 1.0.$$

(In both cases, we use an amplification factor of 1.0, since it would be unconservative to modify the amplification factor to be less than the sum of 1.0.)

Applying the preceding data to AISC Eqs. (H1–1) and (H1–2),

$$.257 + [1.0(4.93 \text{ ksi})/23.76 \text{ ksi}] + [1.0(9.23 \text{ ksi})/27 \text{ ksi}] = .80 < 1.0$$

$$(4.85 \text{ ksi}/21.6 \text{ ksi}) + (4.91 \text{ ksi}/23.76 \text{ ksi}) + (9.23 \text{ ksi}/27 \text{ ksi}) = .774 < 1.0$$

Although both equations work comfortably, the student should perhaps do a few more iterations to make sure there is not a more economical section. (The best choice is a W 12 × 170.)

8.6 Summary

Beam-columns are members under axial load that also are subjected to appreciable bending moments. In a compression member, the deflection caused by the bending moment can be further exaggerated due to the axial load effects. This is referred to as secondary moment. The evaluation and design of beam-columns are handled through the use of interaction equations, which assume a member to be adequate if the ratios of actual to allowable stresses are less than or equal to 1.0.

EXERCISES

1. Explain what is meant by "moment amplification" in a beam-column. Why would a member under axial tension and bending not have this effect?
2. Why are beam-columns more prevalent than a simple column throughout construction? What assumptions are made many times with the design of columns?
3. Using the combined stress formula, calculate the combined stress on the outside of each flange for the eccentrically loaded W 12 × 72 column shown here.

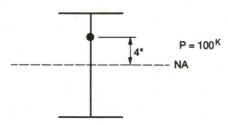

4. Calculate the adequacy of a W 14 × 53 that is 16 feet long with pinned ends. The member is supporting an axial load of 175 kips, a moment about the strong axis of 100 kip-ft, and a moment about the weak axis of 60 kip-ft. The steel is A36, and the member is part of a frame subjected to sidesway.

5. Using the information given in Exercise 4, calculate the reduction of axial load that would make the interaction formulas approximately equal to 1.0 if the weak axis moment were eliminated.

6. A pin-connected W 12 × 120 made of A36 steel is subjected to an axial load of 200 kips and moment M_x = 155 kip-ft. If C_{mx} = 1.0 and the column is 14 feet long, will it be adequate per AISC specifications?

7. Select the most economical W14 section for the beam-column shown in the figure. Assume that the column ends are pinned and that sidesway can occur. The steel is A36.

8. Select the most economical section, using the information as given in Exercise 7, but this time assume the column is braced at its mid-point in the weak direction.

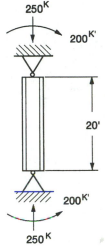

9. Select the most economical W12 section for the beam-column shown in the figure. Assume that the column ends are fixed supports and are restrained against rotation. The end moments are assumed equal to $PL/8$. The steel is A36 and K = .65.

10. Rework Exercise 9, assuming that the column length, l, is changed to 22 ft and the ends of the column are unrestrained.

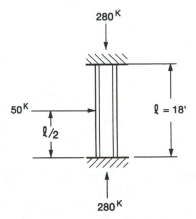

9

BOLTED AND RIVETED CONNECTIONS

9.1   Introduction and Historical Perspective

The need to join steel members together has existed since the introduction of steel as a building material in the latter half of the nineteenth century. Two of the most common methods of connecting steel structures—high-strength bolting and riveting—will be discussed in this chapter, and welding will be the topic of Chapter 10.

An early method of connecting steel members that was widely accepted was riveting. Riveting consisted of heating a "slug" of steel to approximately 1800°F and inserting this slug into holes that joined the members together. Once in the holes, the ends of the soft slug were formed by using a pneumatic hammer to shape the ends of the rivet and to fill up more of the original hole. Many end shapes for rivets were utilized, as shown in Figure 9–1, but by far the most common was the "buttonhead" shape.

Rivets have virtually been replaced in today's world with the advent of high-strength bolting, due primarily to economic considerations. Riveting is a labor-intensive operation requiring a crew of approximately four to five skilled people and is a slower process in terms of installation. High-strength bolting is much quicker, requiring a smaller, less-skilled, two-person crew. Although riveting is rarely used today, many existing structures (approximately pre-1960) are made with rivets; therefore, this method should still be considered.

High-strength bolting is one of the most common procedures used today in the connection of structural steel members. High-strength bolts are made from medium-carbon or alloy steel and have very large tensile strengths. The use of high-

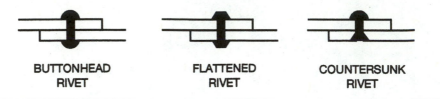

BUTTONHEAD
RIVET

FLATTENED
RIVET

COUNTERSUNK
RIVET

Figure 9–1   Types of Rivets.

strength bolting began in earnest during the 1950s and, by the early 1960s, had virtually replaced riveting altogether. Besides the economic advantages that bolting enjoys, other advantages, such as higher fatigue strength and easier retrofitting ability, are also important.

The two most common high-strength bolts are the A325 bolt and the A490 bolt. The A325 bolt is the most common high-strength bolt used today and is made from heat-treated medium-carbon steel, while the A490 bolt is a higher-strength bolt manufactured from an alloy steel for situations requiring its improved properties.

The proper installation of bolts is based on achieving adequate clamping forces between the members being joined. This clamping force is provided by attaining the proper tension in the bolts, which is required to prevent slippage of the plates being connected and to prevent the bolts from becoming loose. This is especially critical in structures that are subject to repeated stresses or stress reversals. Proper bolt tension is accomplished by using one of the four accepted methods outlined by the AISC's *Specification for Structural Joints Using ASTM A325 or A490 Bolts*. These methods are briefly outlined next.

Turn-of-the Nut Method

This method installs bolts by first placing them in a snug-tight condition and then turning the nut no more than one full turn.

Calibrated Wrench Method

This method utilizes an automatic wrench calibrated to "stall" at a certain torque, producing the required tension. Proper calibration and maintenance of the wrench is critical in achieving the desired results.

Direct Tension Indicators (DTIs)

This method installs bolts using a washer with arched protrusions placed on its bottom surface. As the bolt is tightened, these protrusions become flattened, and by measuring the gap between the bolt head and the plate, one can assess the correct tension placed on the bolt (Figure 9–2). Research[1] has shown that DTIs are

A) Direct Tension Indicator (DTI).

B) DTI before tensioning.

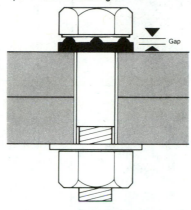

C) DTI after tensioning.

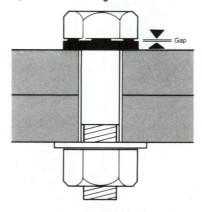

Figure 9–2   Direct Tension Indicators. (Courtesy of J & M Turner Inc.)

Figure 9–3 Installation Procedure of Calibrated Bolt Assemblage Known as the Rapid Tension Bolt. (Courtesy of NSS Industries.)

very reliable in achieving proper bolt tension and effectively inhibit the loss of pretension that may occur over extended periods of time.[2]

Calibrated Bolt Assemblies

This method utilizes a specially calibrated assembly of bolts, nuts, and washers to estimate the proper tension in the bolt. In this method, splined bolts are tightened with a special wrench, causing the tip of the bolt to shear off once the proper tension has been achieved (Figure 9–3).

9.2 Types of Connections

No matter which types of steel shapes are utilized, the connection of steel members always boils down to a "plate-to-plate" configuration. Therefore, the two most common configurations for connections are the "two-plate" and "three-plate" systems, sometimes referred to as the lap joint and butt joint (see Figure 9–4).

Although two-plate and three-plate systems might seem simplistic, their use is widespread throughout steel construction. Studying the typical beam-to-column connection shown in Figure 9–5, we notice that the angle-to-column flange connection is a two-plate system and that the angle-to-beam web connection is a three-plate system.

One of the most important distinctions in any connection system is the number of shear planes passing through a single bolt or rivet. In the two-plate system, the student will notice that as the plates are pulled, movement of the connection plates will attempt to rip the bolts apart along one shear plane should slippage

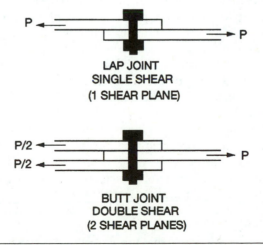

LAP JOINT
SINGLE SHEAR
(1 SHEAR PLANE)

BUTT JOINT
DOUBLE SHEAR
(2 SHEAR PLANES)

Figure 9–4   Common Types of Bolted and Riveted Connections.

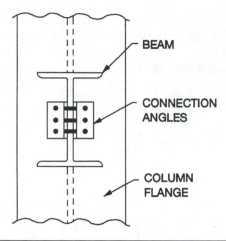

Figure 9–5   Typical Beam-Column Connection.

occur. This phenomenon is referred to as *single shear*. Consequently, in the three-plate system, the bolts are trying to be ripped apart along two shear planes. This phenomenon is referred to as *double shear*. The most important point of difference between single and double shear is the number of areas per bolt that are resisting stress. This will be a major focus when analyzing and designing bolted connections later in this chapter.

Another category under the heading of connection types revolves around the performance of the connection under loading. A connection can be designed as a **bearing-type connection** or a **slip-critical-type connection**. The basic difference between the two types is an assumption of slippage under service load, which results in the use of different allowable shear stresses.

The slip-critical-type connection is assumed to be slip-free under all service load conditions and is to transfer load through the joint by friction between the connected plates. This type of connection is used for structures that have high-impact load conditions or where slippage in the joint is considered undesirable by the designer.

The bearing-type connection is assumed to slip under very high load conditions. If this slippage occurred, the joint would transfer load through shear on the bolts and plate bearing. This type of connection is used when the designer feels the structure is less susceptible to impact, stress-reversals, or vibration.

The difference in the allowable shear stresses for these two types of connections can be found in Table 9–1, which is reproduced from Table J3.2 in the AISC specification.

The reader will notice that the allowable stresses for bearing-type connections are substantially higher than those for the slip-critical-type connection. This

Table 9–1   Allowable Stress on Different Fasteners per the AISC Specification. (Courtesy of the American Institute of Steel Construction, Inc.)

Description of Fasteners	Allowable Tension[g] (F_t)	Allowable Shear[g] (F_v)					Bearing-type Connections[i]
		Slip-critical Connections[e,i]					
		Standard size Holes	Oversized and Short-slotted Holes	Long-slotted holes			
				Transverse[j] Load	Parallel[j] Load		
A502, Gr. 1, hot-driven rivets	23.0[a]						17.5[f]
A502, Gr. 2 and 3, hot-driven rivets	29.0[a]						22.0[f]
A307 bolts	20.0[a]						10.0[b,f]
Threaded parts meeting the requirements of Sects. A3.1 and A3.4 and A449 bolts meeting the requirements of Sect. A3.4, when threads are not excluded from shear planes	$0.33F_u$[a,c,h]						$0.17F_u$[h]
Threaded parts meeting the requirements of Sects. A3.1 and A3.4, and A449 bolts meeting the requirements of Sect. A3.4, when threads are excluded from shear planes	$0.33F_u$[a,h]						$0.22F_u$[h]
A325 bolts, when threads are not excluded from shear planes	44.0[d]	17.0	15.0	12.0	10.0		21.0[f]
A325 bolts, when threads are excluded from shear planes	44.0[d]	17.0	15.0	12.0	10.0		30.0[f]
A490 bolts, when threads are not excluded from shear planes	54.0[d]	21.0	18.0	15.0	13.0		28.0[f]
A490 bolts, when threads are excluded from shear planes	54.0[d]	21.0	18.0	15.0	13.0		40.0[f]

[a]Static loading only.

[b]Threads permitted in shear planes.

[c]The tensile capacity of the threaded portion of an upset rod, based upon the cross-sectional area at its major thread diameter A_b shall be larger than the nominal body area of the rod before upsetting times $0.60F_y$.

[d]For A325 and A490 bolts subject to tensile fatigue loading, see Appendix K4.3.

[e]Class A (slip coefficient 0.33). Clean mill scale and blast-cleaned surfaces with Class A coatings. When specified by the designer, the allowable shear stress, F_v, for slip-critical connections having special faying surface conditions may be increased to the applicable value given in the RCSC Specification.

[f]When bearing-type connections used to splice tension members have a fastener pattern whose length, measured parallel to the line of force, exceeds 50 in., tabulated values shall be reduced by 20%.

[g]See Sect. A5.2

[h]See Table 2, Numerical Values Section for values for specific ASTM steel specifications.

[i]For limitations on use of oversized and slotted holes, see Sect. J3.2.

[j]Direction of load application relative to long axis of slot.

higher allowable stress will translate into fewer bolts needed per connection when designing with a bearing type of joint. The bearing type of connection would be used for structures that are not likely to be loaded in a sudden, high-stress manner, and the friction between the plates will remain as the primary load transfer mechanism until slippage occurs. In structures subjected to high-impact stresses, the lower allowable stress translates into more bolts per connection. This increased number of fasteners per connection provides a safeguard against slippage.

Before closing this section, it should be noted that slip-critical connections and bearing connections should be installed differently. Specifications[3] require that slip-critical connections be "fully tensioned" to 70% of the bolt's minimum tensile strength, while bearing-type connections need to be only installed in a "snug-tight" condition. The reality of steel construction is that many high-strength bolted connections are installed identically, whether they are friction-type or bearing-type. Therefore, we must realize that it is very difficult and time-consuming to distinguish between correct and incorrect tensioning after installation. This underscores the importance of proper installation techniques and shows that proper bolt tensioning is paramount to the structure's economic, as well as structural, well-being.

9.3   Modes of Failure in Bolted Connections

The possible modes of failure in the vicinity of bolted or riveted connections are shown in Figure 9–6 and are listed as follows:

- Shearing of Bolts.   Bolts break due to excessive shear forces along predetermined shearing planes.
- Plate crushing by bolt.   Commonly referred to as bearing failure where the bolt comes into contact with the edge of the bolt hole and causes a crushing type of failure in the plate.
- Tear-out failure of plate.   Ripping failure in the plate due to excessive shear forces can be common if the plate is thin or bolts are close to the edge of the plate.
- Plate failure across net area.   This type of fracture failure was referred to in tension member design. (The last mode of failure will not be discussed here, since a previous discussion in Chapter 4 should have sufficiently covered this topic.)

The primary mode of failure in bolted connections (and therefore the usual controlling feature of connection design) is bolt shear. The shear stresses in this case are assumed split evenly among all bolts in the connection, because we assume the plate to be rigid and nondeforming. This is the standard assumption by the average designer, although the approach is oversimplified, because the typical connection is flexible enough to allow unequal deformations resulting in lower stresses on the bolts in the center of the connection area.[4]

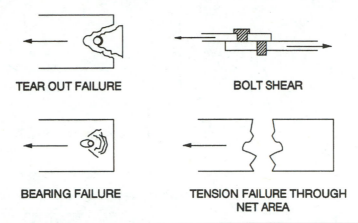

TEAR OUT FAILURE   BOLT SHEAR

BEARING FAILURE   TENSION FAILURE THROUGH
                              NET AREA

Figure 9–6   Common Potential Failures in Bolted Connections.

The shear stress on a bolt can be calculated using the direct stress formula, $f = P/A$. P is the bolt's share of the load to be resisted, and A is the cross-sectional area of the bolt. The student at this point should be reminded of the single and double shear cases because in the case of double shear, the bolt has twice as much area resisting the load. The allowable shear stresses for the A325 and A490 bolts were shown in Table 9–1, or Table J3.2 in the AISC specification.

The second mode of failure is the crushing of the plate by the bolt bearing on it. This failure mechanism is relatively rare unless very thin plates are used. Still, it must be checked as part of any analysis or design. The bearing stress of a bolt on a plate is given again by the direct stress formula, $f = P/A$, the only difference being that the resisting area is the projection of the contact area of the bolt on the plate. This area is simply the thickness of the plate multiplied by the diameter of the bolt (see Figure 9–7).

The allowable bearing stress, if deformation is not a consideration and adequate bolt spacing and edge distance requirements are met, is $1.5\,F_u$, where F_u is the specified minimum ultimate tensile strength of the connected parts. For standard holes where there are more than two bolts in the line of stress (and where deformation is to be considered), the allowable bearing stress is given as $1.2\,F_u$. For other cases regarding bearing, the allowable stress requirements can be found in Section J3.7 of the AISC specification.

The last mode of failure is plate tearing or shearing, which is probable only when bolt spacings and edge distances become very small. To effectively eliminate this behavior, the AISC lists minimum spacing and edge distance requirements in Sections J3.8 and J3.9. Generally, if the allowable bearing stress is taken at $1.2\,F_u$, the minimum bolt spacing (center-to-center of bolts) should be taken as $3d$ (where d is the bolt diameter) and the minimum edge spacing should be taken as $1.5d$. When using an allowable bearing stress in excess of $1.2\,F_u$, the aforementioned

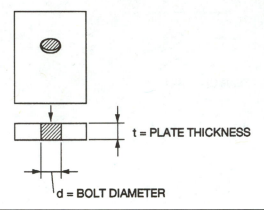

Figure 9–7   Bearing Area of Bolt in Bolt Hole.

sections in the AISC specification should be consulted to determine the minimum bolt and edge spacings.

9.4   AISC Specifications for Bolted and Riveted Connections

The allowable stress philosophy of steel design strives to keep the actual stresses in a connection below some specified value. These specified values are as previously shown in Table 9–1 or as listed in Section J3 in the AISC *Manual of Steel Construction*. The primary values for allowable stress against which the connection's actual stress should be checked are those for **bolt shear** and **bearing on the members being fastened**. Although typically not of great concern, minimum bolt spacing and edge distance might also be checked so that a failure due to plate tearing does not occur. The tension failure mechanism across the net area should be checked in the tensile capacity calculations during some part of the total design.

The allowable stress values for bolt shear were previously listed in Table 9–1 of this chapter. The student should remember that the area resisting this shear force is the cross-sectional area of the bolt and that a bolt may have more than one area resisting (as in the case of double shear).

The allowable stress values for plate bearing were discussed in the previous section. The area that resists this bearing force is the projected area of the bolt on the plate which it bears. This projected area is conveniently taken as the bolt diameter multiplied by the thickness of the plate.

9.5   Design of Bolted and Riveted Connections

As discussed previously, the designer encounters two types of problems. The first is the "evaluation" problem, for which an existing connection is given and its capacity for a new load condition is needed. The known entities in this problem are load, area, and allowable stress. By using the direct stress equation, the designer calculates either the actual stress or the allowable load. A comparison of actual stress to the allowable stress or the allowable load to the actual load would then determine the adequacy of the connection.

The second type of problem is the "design" problem, for which the designer would find the minimum area of bolts and plate bearing area needed to make the connection work. This type of problem is more open-ended, since a number of solutions may be possible. The known quantities in this problem are load and allowable stress, and we rearrange the direct stress equation to solve for area required. The designer must keep in mind that determination of the area required in a connection will also impact other parts of the total design, namely, net area and bolt layout. All factors must be considered in the complete design.

The following examples present solution techniques for these problem types.

EXAMPLE 9.1

Determine the adequacy of the connection shown in the figure. The bolts are A325 with threads not excluded from the shear plane (A325-N), and the steel plates are A36. The applied load is 75 kips, and the connection is bearing-type.

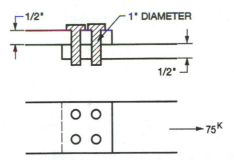

Remember that the two items to check concerning the connection are shear and bearing, as long as the bolt spacing and edge layout are adequate. (In our problems we will assume that they are adequate.) This is an evaluation type of problem, which we will evaluate by comparing actual load to the allowable load.

Shear

In the two-plate, single-shear condition we have one shear plane in each bolt. Therefore:

$$A = \pi/4(1 \text{ in.})^2 \times 4 \text{ bolts} = 3.14 \text{ in.}^2$$

Allowable stress = 21 ksi   (per Table 9–1 for A325 bolts with the threads in the shear plane, bearing-type connection)

Calculate the allowable load:

P_{allow} = 3.14 in.^{2} × 21 ksi = 65.9 kips

Compare this to the actual load:

65.9 kips < 75 kips   No Good.

Bearing

Check the bearing (always) even though shear failed. Area of bearing surface:

A = 1/2 in. × 1 in. × 4 bolts = 2 in.^{2}

Calculate the allowable stress in the bearing (considering no deformation and spacings are adequate):

Allowable bearing stress = 1.5(58 ksi) = 87 ksi

P_{allow} = 2 in.^{2} × 87 ksi = 174 kips

174 kips > 75 kips   OK, but connection still fails.

(Remember that as part of a complete evaluation, we would also check the plate tension over the net and gross area for their adequacy.)

EXAMPLE 9.2

The connection to be made in the lap joint shown in the accompanying figure is to withstand a tensile force of 100 kips. If the connection is a slip-critical type and the steel is A36 and bolts are A325-X (threads not in the shear plane), design the bolt size and number of bolts to be used.

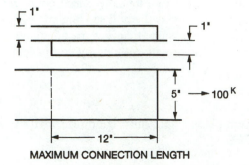

MAXIMUM CONNECTION LENGTH

Shear

Calculate bolt area needed based on shear requirements:

$P = 100$ kips

Allowable shear stress = 17.0 ksi

$A_{req'd} = 100$ kips/17.0 ksi = 5.88 in.^{2} (Try 1-in.-diameter bolts.)

Using 1-in.-diameter bolts ($A = .7854$ in.^{2}), you would need 5.88 in.^{2}/(.7854 in.^{2}/bolt) = 7.49 bolts, or 8 bolts. This choice is probably a wise one because you could lay out the bolts as shown in the next figure, which maintains adequate bolt spacing. A smaller diameter of bolt would lead to a larger number of bolts, and a minimum spacing of bolts might be in jeopardy.

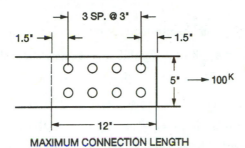

MAXIMUM CONNECTION LENGTH

Bearing

Check the bearing area using 1-in. bolts to ensure its adequacy:

$P = 100$ kips

Allowable bearing stress = 1.5 F_u = 1.5(58 ksi) = 87 ksi

$A_{req'd} = 100$ kips/87 ksi = 1.15 in.^{2}

Area with eight 1-in. bolts = 8 × 1 in. diameter × 1 in. thickness = 8 in.^{2}
  OK, since actual area is larger than required.

(Again, in an actual design we would need to check tension member's net area and gross area capacities.)

EXAMPLE 9.3

Calculate the adequacy of the butt joint shown in the figure, if the bolts are 5/8-in.-diameter A490's with the threads not in the shear plane. The steel is A36, and the connection is a bearing type.

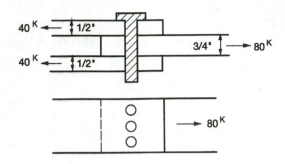

Shear

Check the bolt shear adequacy by calculating the bolt shear stress:

$A = \pi/4(5/8 \text{ in.})^2 = .3068 \text{ in.}^2$ per bolt × 3 bolts = .92 in.^{2}

Remember that you have double shear. Therefore, $A = .92 \times 2 = 1.84 \text{ in.}^2$

Actual bolt shear stress = 80 kips/1.84 in.^{2} = 43.47 ksi

Allowable shear stress from Table 9–1 = 40 ksi. Therefore, since actual stress is greater than allowable stress, the connection fails per shear requirements.

Bearing

Check the bearing (even though connection failed in shear):

Bearing area = 3 × 5/8 in. diameter × 3/4 in. thickness = 1.406 in.^{2}

Actual bearing stress = 80 kips/1.406 in.^{2} = 56.9 ksi

which is less than the allowable of 1.5 F_u (87 ksi). Therefore, bearing stresses are adequate although connection still fails in shear.

(Notice that in the bearing calculation we used the 3/4-in. plate because the critical bearing area in a butt joint will be the smallest area subjected to the full load. Therefore, the 3/4-in. plate under a load of 80 kips was more critical than two 1/2-in. plates under the full 80-kip load.)

In closing, it should be noted that the AISC manual, Part 4, contains a number of tables that provide the designer a quick and effective way to select the proper number of bolts and angle sizes for standard beam connections. Although not covered in this text (because students must first understand the fundamentals in bolted connections), these tables are convenient and should be explored by students planning to continue with structural steel design.

9.6   Summary

High-strength bolting is the most common method of connecting steel members used in construction today. It is very important that proper bolt tension be achieved during the installation process, for which a number of different methods strive to ensure this outcome.

Analysis and design of bolted connections focus primarily on preventing bolt shear and plate bearing failures, although other mechanisms (such as tensile modes of failure) must always be checked. The good designer will realize that changes to the connection will most probably be reflected in other parts of the member design, such as the tensile capacity.

EXERCISES

1. Discuss the significance of a slip-critical-type versus a bearing-type connection. Where would you expect to see each of these connections used? Are they both installed in similar fashion?
2. Discuss the direct tension indicators used in the tensioning of bolts. Do you think this is a reliable method of obtaining the proper tension? Do you believe imperfections such as burrs or rust may affect this method?
3. Compare the calibrated bolt assemblage method of bolt tensioning to the method discussed in Exercise 2. How does this method compare?
4. Determine the allowable load on the bearing connection shown in the figure, if the bolts are A325-N, 7/8-in. diameter and the steel is A36. Check the plate tension also.

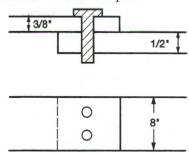

5. Check the adequacy of the slip-critical connection shown in the figure, if the bolts are A490-X, 3/4-in. diameter and the steel is A36. Check the plate tension also.

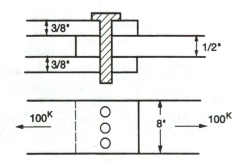

6. Design the size and number of A325-X bolts required in the bearing connection shown in the figure. Check shear and bearing, assuming the required edge distance is 1-1/2 in. to bolt center and that the minimum spacing between centers of bolts is 3d. The steel is A36.

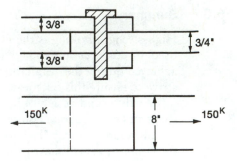

7. Redesign the connection in Exercise 6, if the two outside plates are 1/4 in. thick, and consider deformation around the holes.

8. Recalculate the adequacy of the connection shown in Exercise 5, if the connection is now a bearing type. How does this change your answer?

9. In a bearing-type connection, why are the allowable shear stresses given in Table 9–1 so much lower for bolts with threads included in the shear plane? Why don't these stresses change when a slip-critical connection is considered?

10. Design the proper diameter for the bearing-type lap joint shown in the accompanying figure. The bolts are A490's with the threads not in the shear plane, and the steel is A36. Assume proper spacing and edge distance and that deformation is not a concern.

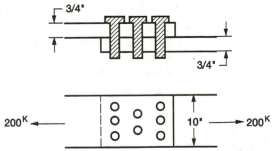

REFERENCES

1. John A. Struik, Abayomi O. Oyeledun, and John W. Fisher, "Bolt Tension Control with a Direct Tension Indicator," *Engineering Journal*, AISC, 1973.

2. J.O. Surtees and M.E. Ibrahim, "Load Indicating Washers," *Civil Engineering*, ASCE, April 1982.

3. *Specifications for Structural Joints Using ASTM A325 or A490 Bolts*, Research Council on Structural Connections of the Engineering Foundation, June 8, 1988.

4. Jack C. McCormac, *Structural Steel Design: ASD Method*, 4th ed. (New York: Harper Collins Publishers, 1992), p. 308.

10

WELDED CONNECTIONS

10.1   Introduction

Welding can be thought of as the fusing of two pieces of metal together to form a continuous, rigid plate. The earliest welding was probably accomplished by craftsmen and artists in ancient times. Of the many modern-day welding techniques, the two basic categories are gas and arc welding. These modern techniques can trace their roots to the late nineteenth century when arc welding was first patented and used on a limited scale.

Generally, all welding processes will have the following common elements:

- Base metals
- Heat source
- Electrode or welding rod
- Shielding mechanism

The base metals are simply the pieces to be joined together. They are joined with some type of heat source. In structural steel welding, this heat source is most frequently generated through an electric arc, in gas welding, through the burning of gas (typically acetylene and oxygen) at the end of a welder's "torch." In both cases, the heat source melts not only the base metal but also an electrode or welding rod. As the electrode or welding rod is melted, it deposits additional steel into the area of the weld. The electrode or welding rod may be thought of as the "weld metal."

The weld must be protected from contact with the surrounding air during its cooling period. This is usually accomplished through some type of shielding mechanism, which includes a gaseous cloud or immersion of the electrode into a material generally referred to as flux. If air is allowed to penetrate the weld during its cooling period, it can greatly reduce the strength and quality of the weld due to pitting.

The use of welding is very popular in steel construction because it has a number of advantages. Among them are the savings produced by the reduction of splicing plates, the ease of welding odd shapes (i.e., pipes), and the ease of implementing field changes. Disadvantages of welding include their fatigue behavior and quality assurance. The former of these disadvantages has led to the development of fatigue criteria and special details in practically all welded structures.

As mentioned earlier, the most common method of welding structural steel is arc welding. This chapter will consider the two most basic types of arc welding: the shielded metal arc welding process and the submerged arc welding process (Figure 10–1).

The shielded metal arc welding (SMAW) process, the most traditional type, is

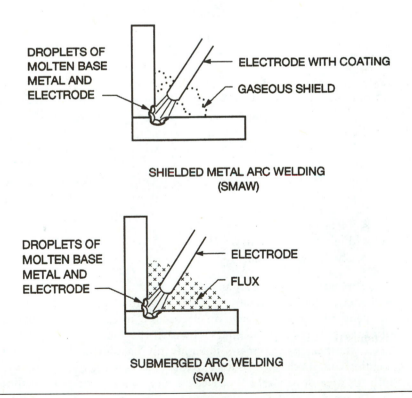

Figure 10–1   Illustration of Common Welding Techniques.

manually produced. The generation of heat is from an electric arc, and the electrode, usually designated by a term such as E70XX, is melted into the weld area to fuse together with the base metal. In the aforementioned electrode designation, E70XX, the 70 represents the ultimate tensile strength (ksi) of the electrode. The postscript symbols may reflect a variety of characteristics such as coating and positions.

Submerged arc welding (SAW) also uses an electric arc for its heat source. This method is used most often in a fabrication shop and is accomplished by an automated welding machine. The machine will lay down a coating of a granular material, called flux, that shields the weld. This type of weld is considered to be superior because of the uniform quality characteristics and mechanical properties.

10.2   Common Types of Welds

There are many different ways to classify welds based on the type of weld, the welding position, and the type of end treatment in which a plate may be fabricated for better adaptation to the welding procedure. The standard welding symbols as given by the AISC are shown in Figure 10–2. In this section we will be most concerned with identifying the characteristics and terminology for the common weld types.

The most common type of weld, the fillet weld, comprises approximately 85% of all welds that are produced. These welds are used to join two pieces of steel, forming a perpendicular corner where the weld is placed (Figure 10–3). The abundance of fillet welds is due in part to their ease of production because of the "pocket" formed by the perpendicular edges. This pocket serves as a holder of the molten steel and eliminates the need for additional backup plates.

The most common fillet weld is the equal leg fillet weld, where the leg dimensions are the same length. The different parts of the fillet weld are shown in Figure 10–4. The most important part of the fillet weld from a design standpoint is the throat dimension. The throat dimension is the shortest length from the root of the weld to its face. This distance is critical because it is the probable line of failure through the weld. In an equal leg fillet weld this theoretical throat distance will be *.707 × the leg*. Fillet welds are called out by their leg size in equal leg welds (i.e., a 3/8 fillet weld has a leg 3/8 in. long).

The other common type of weld is referred to as a groove or butt weld. This weld is used when connecting two plates that lie in the same plane (Figure 10–3). A groove weld is referred to as a **full penetration groove weld** if the weld extends the full thickness of the plate being joined, and as a **partial penetration groove weld** if the weld does not extend the full thickness. Besides this difference, a groove weld has many other variations based on the depth of the weld, the edge fabrication of the plates, and the use of additional weld material. In any case, groove welds are more difficult to produce not only because of additional

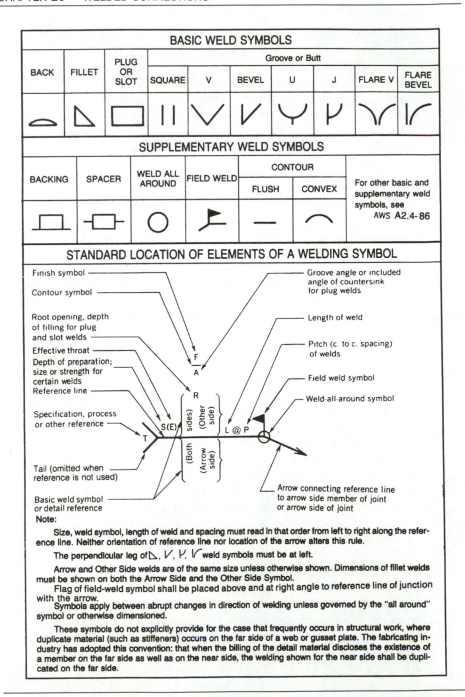

Figure 10–2 Standard Welding Symbols. (Courtesy of the American Institute of Steel Construction, Inc.)

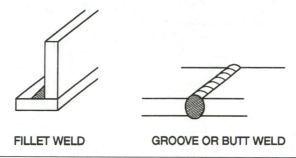

FILLET WELD   **GROOVE OR BUTT WELD**

Figure 10–3   Schematic Illustration of Fillet and Groove Welds.

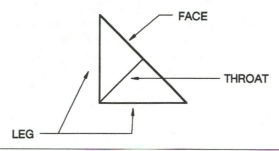

Figure 10–4   Standard Terminology for a Fillet Weld.

plate fabrication but also because the weld is not confined in a premade "pocket" such as the fillet weld.

The critical throat distance in a full penetration groove weld is taken by the AISC to be the thickness of the thinnest plate joined. That is, if a 1/4-in.-thick plate were groove welded to a 1/2-in.-thick plate, the throat distance of the groove weld would be 1/4 in.

10.3   Potential Modes of Failure in Welded Connections

There are basically two manners in which a welded connection can fail: Either the weld itself cracks under some type of stress, or the base material being joined actually cracks with the weld intact. Both of these mechanisms are induced due to the shear stresses that act along some plane, either a plane in the weld itself or a plane in the base metal. This plane of failure in the weld is referred to as the **effective weld area.**

The effective weld area is that area most susceptible to cracking, assuming the weld to be of uniform quality. This area would be the smallest area resisting the loads placed upon the weld and is therefore the most likely to fail. Whether these loads are placing the weld in tension, compression, or shear, the weld will tend

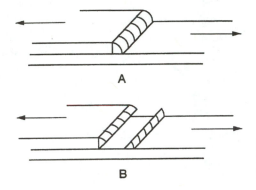

A

B

Figure 10–5 Typical Failure Plane in a Welded Connection Exposing the Effective Area. (a) Weld intact. (b) Weld broken.

toward resisting a stress with this effective area. We can best picture this area as the surface area that would remain *after* the weld had failed. This effective area is simply the critical throat distance (sometimes referred to as the effective throat) multiplied by the length of the weld (Figure 10–5).

As stated earlier, the critical throat distance for an equal leg fillet weld is .707 × leg. This makes the effective weld area for an equal leg fillet weld, .707 × leg × length of weld. Thus, for a groove weld (which had a critical throat distance equal to the thickness of the thinner part), the effective area would be the critical throat times the weld length.

The effective areas, as already described, are to be used when welding by the shielded metal arc welding (SMAW) process; but since the submerged arc welding (SAW) process leads to an inherently better fillet weld, the AISC specification permits a modified critical throat distance. For a fillet weld made by the SAW process, the following effective weld areas are proposed by the AISC:

- For fillet welds 3/8 in. and smaller:

    Effective area = Leg × Length of weld

- For fillet welds larger than 3/8 in.:

    Effective area = (Leg + .11 in.) × Length of weld

To gain further insight to the requirements of the AISC specification regarding welding, students should familiarize themselves with Section J2 (page 5–65) of the AISC specification. These requirements are especially important with regard to maximum and minimum welds sizes used in proper connection design.

10.4 AISC Allowable Stresses in Welded Connections

The allowable stress criteria as established in the AISC specification are very straightforward. Remember that both the weld itself and the plates they are connecting should both be checked to determine completely the connection's capacity.

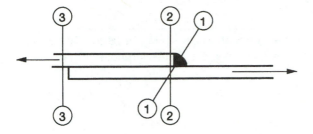

Figure 10–6 Potential Failure Zones in a Welded Connection. (1–1) Shear through weld. (2–2) Shear through plate. (3–3) Tension in plate.

The planes of failure in a typical welded connection can be summarized in Figure 10–6.

The transfer of stress through fillet welds is assumed to be shear over the effective weld area, and failure is assumed to occur in plane 1–1, as shown in Figure 10–6. When subjected to a shearing force perpendicular to the axis of the weld, the allowable shear stress for a fillet weld is given in Table J2.5 (page 5–70) of AISC specification as $0.30\ F_u$, where F_u is the ultimate tensile strength of the electrode being used. When a fillet weld is subjected to tension or compression parallel to the weld's longitudinal axis, the allowable stress in the base metal must not exceed previously defined limits (which would be $0.60\ F_y$ in tension and $0.40\ F_y$ in shear).

The stress in full penetration groove welds is transferred in exactly the same manner as it is carried in the plates. The allowable shear stress on the effective area of the groove weld is again $0.30\ F_u$. The allowable stress for tension and compression on the welds, either parallel or perpendicular to the weld's axis, is the same as in the base metal (previously stated). However, when tension stress is applied normal to the weld's effective area, the AISC specification recommends that "matching" weld metal—that is, electrode material with mechanical properties that meet or exceed that of the weakest base metal—be used. A complete table outlining this can be found in the American Welding Society's *Welding Handbook*.[1]

10.5   Design of Welded Connections

Welds are used to transfer forces from one piece of steel to another. This transfer of force produces stress on the welds, which can be quantified in terms of the direct stress equation that we have repeatedly used, $f = P/A$. The area is now the effective weld area that was discussed in a previous section.

Weld analysis or design can be viewed in terms of one-inch segments of the weld under consideration. This is because a one-inch segment of a weld has a particular load capacity based on the strength of the electrode used and the size of the weld. To double this load capacity, the designer merely needs to double the weld length.

Again, two basic problems are commonly encountered: evaluation and design. The following two examples illustrate solution techniques to these common problems.

EXAMPLE 10.1

Determine the allowable load on the connection shown in the figure. The welds are 5/16 in. fillet with E70XX electrodes using the SMAW process. Plates are made of A36 steel.

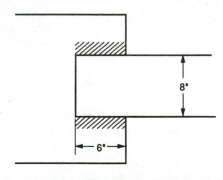

NOTE:
BOTH PLATES ARE 1/2" THICK

Since the problem is to determine the allowable load on the connection, it is an evaluation type of problem. Determine the effective weld area and calculate the AISC allowable stress:

Effective weld area = $(5/16)(.707) \times (6 \text{ in.} + 6 \text{ in.}) = 2.65 \text{ in.}^2$

Allowable stress = $0.30 \ F_u = .30(70 \text{ ksi}) = 21 \text{ ksi}$

Allowable load = $2.65 \text{ in.}^2 \times 21 \text{ ksi} = 55.6 \text{ kips}$

Also check tension and shear on base plates to make sure they are not weaker than the weld itself.

Plate Tension

$A_{gross} = 4 \text{ in.}^2$

Allowable stress = $0.60 \ F_y = .60(36 \text{ ksi}) = 21.6 \text{ ksi}$

Allowable load = $4 \text{ in.}^2 \times 21.6 \text{ ksi} = 86.4 \text{ kips}$

Plate Shear (around weld area)

Area = $12 \text{ in.} \times 1/2 \text{ in.} = 6 \text{ in.}^2$

Allowable stress = $0.40 \ F_y = .40(36 \text{ ksi}) = 14.4 \text{ ksi}$

Allowable load = $6 \text{ in.}^2 \times 14.4 \text{ ksi} = 86.4 \text{ kips}$

Therefore, the weld capacity (equal to 55.6 kips) controls the connection's strength.

(*Note:* In actual weld design, the selected weld size should be checked against the minimum and maximum weld sizes that are allowed. For fillet welds, the maximum weld size is 1/16 in. less than the thickness of the plate for plate thickness over 1/4 in., and the thickness of the plate for plates 1/4 in. or smaller. In this example, the maximum fillet weld size would be 7/16 in. (1/2 in. − 1/16 in.). The minimum fillet weld size is given in Table J2.4 (page 5–67) of the specification as 3/16 in. for a 1/2 in. plate. Since our weld is 5/16 in., it fits between the minimum and maximum specified weld sizes and is therefore okay. Students are urged to become familiar with such requirements as they further their knowledge of steel design.)

EXAMPLE 10.2

Design the welds for the connection shown in the figure using two equal length side fillet welds. Use E70XX electrodes with SMAW process and A36 steel.

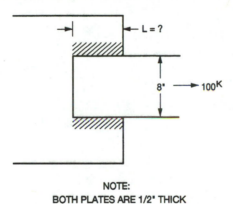

NOTE:
BOTH PLATES ARE 1/2" THICK

This is a design type problem where we are left to design the size and the length of weld to be used. Again, maximum and minimum fillet weld sizes may be used as a starting point when deciding on a course of action. (Students not familiar with maximum weld sizes are referred to Section J2 of the AISC specification for the logic used here.)

Since the plates are each 1/2 in. thick, the maximum size weld per AISC specification is 7/16 in., but let's use 3/8 in. because it will be easier to manually produce. (As in the previous example, when using 1/2 in. plates, the minimum fillet weld size would be 3/16 in.)

Effective weld area = (.707 × 3/8) × length = .265 × length

Since the load is given at 100 kips and the allowable weld stress from the specification is 0.30 F_u, the length can be found as follows:

$A_{\text{req'd}} = P/.30(F_u)$

$(.265)$ Length $= 100$ kips$/.30(70$ ksi$)$

Length $= 17.96$ in.,   use 2 welds 9 in. each

Remember to check tension and shear on plates.

Plate Tension

$A_g = 8$ in. $\times$ 1/2 in. $= 4$ in.^{2}

$F_t = 0.60\ F_y = 21.6$ ksi

Therefore tension capacity of the connection is only:

4 in.$^2 \times 21.6$ ksi $= 86.4$ kips

which is less than what the welds are designed to carry. This would call for the designer to redesign the connection based on plate tension.

Plate Shear

$A = 1/2$ in. $\times$ (18 in.) $= 9$ in.^{2}

$F_v = 0.40\ F_y = 14.4$ ksi

Therefore the shear capacity of the plates would be:

9 in.$^2 \times 14.4$ ksi $= 129.6$ kips

which is more than the 100 kips which the connection is to carry.

EXAMPLE 10.3

Calculate the allowable load that the connection in the figure can hold. The full penetration groove weld uses E70XX electrodes with the SMAW process and connects the plates made from A36 steel.

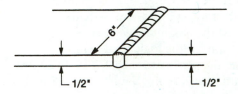

Since the plates are both 1/2 in. thick, the critical throat distance is 1/2 in. The effective weld area is

$A = 1/2$ in. $\times$ 6 in. $= 3$ in.^{2}

The allowable stress for tension perpendicular to the weld's longitudinal axis is $0.60\ F_y$, or $0.60(36\text{ ksi}) = 21.6\text{ ksi}$.

The allowable load based on the weld capacity is:

$$P_{all} = 3\text{ in.}^2 \times 21.6\text{ ksi} = 64.8\text{ kips}$$ (Matching weld metal must be used.)

Checking plate tension, we calculate gross area and multiply that by the allowable tensile stress of the plate:

$$A_g = 1/2\text{ in.} \times 6\text{ in.} = 3\text{ in.}^2$$
$$P_{all} = 3\text{ in.}^2 \times 21.6\text{ ksi} = 64.8\text{ kips}$$

Therefore, the allowable load on the connection is 64.8 kips. (Checking plate shear does not need to be done in this case, since the plane of tension excludes the possibility of shear failure.)

10.6 Fatigue Considerations in Welded Connections

Fatigue in structural members can be defined as the tendency to fail at a lower stress when subjected to cyclical loading than when subjected to a static loading.[2] Although these loadings may cause stresses that are greatly under a material's threshold stress, a member can exhibit cracking due to the large number of repetitions.

Welded structures are extremely sensitive to fatigue cracking because of the discontinuous and flawed nature of welded connections in general. A weld can be thought of as the boundary between two discontinuous plates that binds these plates into a rigid body. Some welded details, such as the cover plate shown in Figure 10–7, are prone to cracking because of the varying rigidity on each side of

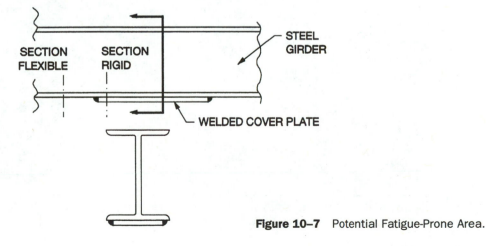

Figure 10–7 Potential Fatigue-Prone Area.

the weld. In this particular case, the weld is solely responsible for transferring the stress from one part of the flange to another over a short distance. This essentially places the weld in a position of extremely high stress.

Besides being located in positions of high stress, welds by their very nature may be the cause of high stress due to the flawed nature of their production. Welds are susceptible to flaws such as the possible inclusion of slag, incomplete fusion, air pockets, improper starts and stops, or arc strikes. Such flaws cause **stress risers**, or areas where the stress concentrates at a much higher rate. The most critical condition for a fatigue crack to initiate is that which combines a flaw with a detail of high stress.[3] These flaws are particularly critical when they are oriented perpendicular to the applied stress.[4] Fatigue crack initiation will be much more likely to occur with a flaw oriented in this manner coupled with a high stress detail. When a crack finally initiates, its rate of growth will depend on the number and intensity of stress cycles. Finally, the member will fracture as the crack size becomes critical. Some member fractures, such as the one shown in Figure 10–8, have been caused by seemingly trivial details such as filling misplaced holes with weld material.

For welded structures subjected to repetitive or cyclical loads, specifications dictate how fatigue-prone details shall be designed. Appendix K of the AISC speci-

Figure 10–8   Crack Caused by Weld-Filled Rivet Hole in a Bridge Girder. (Courtesy of ATLSS Engineering Research Center, Lehigh University.)

Table 10–1   AISC Fatigue Loading Cycle Criteria

Loading Condition	From	To
1	20,000[a]	100,000[b]
2	100,000	500,000[c]
3	500,000	2,000,000[d]
4	Over 2,000,000	

Source: Courtesy of the American Institute of Steel Construction, Inc.
[a]Approximately equivalent to two applications every day for 25 years.
[b]Approximately equivalent to 10 applications every day for 25 years.
[c]Approximately equivalent to 50 applications every day for 25 years.
[d]Approximately equivalent to 200 applications every day for 25 years.

Table 10–2   AISC Allowable Stress Ranges for Various Fatigue Conditions

Category (from Table A–K4.2)	Loading Condition 1	Loading Condition 2	Loading Condition 3	Loading Condition 4
A	63	37	24	24
B	49	29	18	16
B′	39	23	15	12
C	35	21	13	10[a]
D	28	16	10	7
E	22	13	8	5
E′	16	9	6	3
F	15	12	9	8

Source: Courtesy of the American Institute of Steel Construction, Inc.
[a]Flexural stress range of 12 ksi permitted at toe of stiffener welds on flanges.

fication lists different stress categories when dealing with welds on cyclically loaded members. These categories are determined by the type of weld detail and the number of service cycles that a member is expected to incur over its lifetime, as shown in Tables 10–1 and 10–2. As the weld detail becomes more prone to fatigue behavior and as the number of expected cycles in a member's lifetime increases, the AISC specification will reduce the allowable stress on that connection. For a complete description of each fatigue category listed in these tables, the student is referred to Table A–K4.2 (page 5–108 through 5–113) in the AISC specification.

Fatigue behavior in welded connection is a complex and thought-provoking topic in which much of today's research will be reflected in tomorrow's specifications. New methods and new details are constantly being tried and evaluated to try to reduce the problems associated with this behavior. The beginning designer should be made aware of the potential for problems related to fatigue cracking in these details and seek experienced help when involved in this matter.

10.7 Summary

Welding is the fusing of steel plates into a single, rigid member using a heat source, which is typically an electric arc. An electrode or welding rod is melted by the electric arc along with the base metal to form the weld. The most common weld is a fillet weld.

Analysis and design of welded connections utilizes the direct stress formula and allowable stresses as set forth in the AISC specification. The critical area over which the weld is most likely to break is termed the effective weld area. Fatigue behavior in welded connections is a concern in design, and the specifications should be consulted to determine how this behavior is handled.

EXERCISES

1. Discuss the differences between shielded metal arc welding (SMAW) and submerged arc welding (SAW). How does the AISC specification differentiate these two?
2. Why do welds need to be protected from air intrusion while they are being fused? How do we protect these welds from air?
3. Explain what fatigue is and how it can affect welded structural members. How do specifications such as the AISC handle fatigue-related details concerning allowable stresses?
4. Calculate the allowable load that can be held on the welded connection shown in the figure. The weld is a 3/8-in.-fillet weld made from E70XX electrodes. The base steel is A36, and the weld is made by the SMAW process.

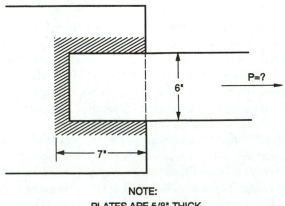

NOTE:
PLATES ARE 5/8" THICK

5. Calculate the length of the full penetration groove weld shown in the figure if the connection is to hold 80 kips. The electrode is an E60XX, and the base steel is A36. The SMAW process is used.

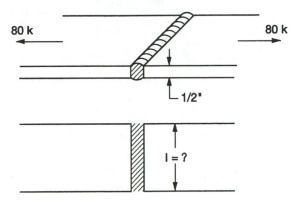

6. If the welded connection shown in the accompanying figure is made from a 1/4-in. fillet weld around the total plate perimeter using the SMAW process, will the connection work? The steel is A242 ($F_y = 50$ ksi, $F_u = 70$ ksi), and electrodes are E70XX.

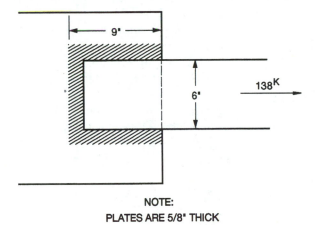

NOTE:
PLATES ARE 5/8" THICK

7. If the connection in Exercise 6 used the SAW process, would the connection's capacity be improved? Explain.
8. Design a fillet weld (size and length) to hold a 200-kip tensile force in the connection shown in the figure. Use side welds initially and an end weld if needed. The electrodes are E80XX, and the steel is A242. Use the SAW process.

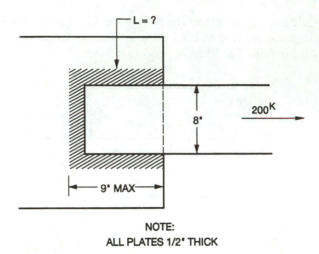

NOTE:
ALL PLATES 1/2" THICK

9. If the connection in Exercise 8 were to use only an end weld and a weld underneath the 8-in. plate, could the connection be adequately designed?

REFERENCES

1. American Welding Society, *Structural Welding Code—Steel* (D1.1–88), 11th ed., Miami, Florida, 1988.
2. U.S. Department of Transportation, Federal Highway Administration, *Bridge Inspector's Training Manual 70,* Washington, D.C., 1979.
3. U.S. Department of Transportation, Federal Highway Administration, *Inspection of Fracture Critical Bridge Members,* Report No. FHWA-IP-86-26, 1986, p. 55.
4. John Fisher, Hans Hausammann, Michael Sullivan, and Alan Pense, Transportation Research Board's National Cooperative Highway Research Program Report No. 206, "Detection and Repair of Fatigue Damage in Highway Bridges," Washington, D.C., June 1979, p. 50.

A

INTRODUCING THE LRFD PHILOSOPHY

A.1 Introduction

Over the last 25 years, the structural design community has been very interested in developing a design procedure that uses a probability-based rationale. With the adoption (1986) of the AISC "Load and Resistance Factor Design Specification," the steel design industry has given its endorsement for such a method. Although Load and Resistance Factor Design (LRFD) has not received the quick and overwhelming acceptance in the design community some would like,[1] few doubt that it will achieve majority usage by designers in the next 15 years or so.

This appendix is a brief summary of the LRFD method of steel design. This summary will try to deliver a simplified explanation of what LRFD is all about and the design procedures for such basic members as beams, columns, and tension members. For a further explanation of this topic, the student is encouraged to read the upcoming text by the author devoted solely to this subject.

A.2 Loads and Structural Resistance

A simplified view of structural mechanics would state that when members are loaded they respond by resisting that load. The failure of a member could occur because the loads that act on a member are larger than those for which it was designed or because the resistance of the member is smaller than anticipated. If we contemplate loads and member resistances, we begin to realize that they are indeed variable over the life of a structure.

As previously discussed in Chapter 2, loads on structures may consist of, among others, dead loads (D), live loads (L), wind loads (W), earthquake loads (E), and snow loads (S). These loads are very different in terms of predictability. The uncertainty of a dead load should be relatively small compared to the variability associated with wind loads or earthquake loads. As designers of structures, we should have a good handle on the major components of dead load such as self-weight, but possibly we would be less certain of the live load requirements to be placed on our structure 20 years into the future. Therefore, we must logically state that live loads are more unpredictable than dead loads. (The reader should remember that in the ASD methodology, a distinction between the type of loads was never discussed as part of the design process.)

The resistance of a member is also variable based on manufacturing, age, and dimensions. When designers specify A36 steel, it is *supposed* to have a minimum yield strength (F_y) equal to 36 ksi. But there is some small percentage of A36 steel members with a yield strength less than 36 ksi. The resistance of some steel members may also decrease with age because of concerns such as corrosion that may reduce the effective area of resistance.

The LRFD method approaches structural steel design using a probabilistic approach, incorporating the variability of loads and member resistances, to arrive at a more rational and (hopefully) cost-effective solution.

A.3   Basic Concepts of Probability, Statistics, and Failure

Probability is the chance or likelihood that a predefined occurrence will actually happen. The frequency of a variable event is the number of times it occurs in a set of observations. Any variable event can be looked at in terms of probability, whether that event is winning the lottery, selecting an ace out of a deck of cards, or having a structure collapse.

All random occurrences will follow some set pattern, relating the number of times an event has occurred with the random variable associated with that event. A curve for any probabilistic event could be defined mathematically, and this is referred to as the probability function.

Many times the probability function for various random events is very complex, and it is desirable therefore to be able to describe the important features of the function. When dealing with the field of probability and statistics, a few basic terms need to be defined. One of the most important of these terms is the **mean**. The mean is the average value (or the value of the random variable with the highest probability of occurring) in a set of observations (Figure A–1). The mean can essentially be thought of as the **expected value** of the random variable, although it will not always be the value that the variable assumes.

Another important description of a probability curve is the spread, or scatter, of the random variable values away from the mean value. This measure of disper-

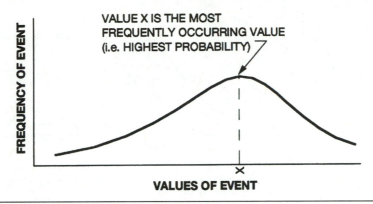

Figure A–1   Typical Probability Function.

sion is termed the **standard deviation,** and one must realize that a probability function can take on many different shapes, which represent the different "spreads," or "scatter," of data about the mean, although they may have the same mean value. Figure A–2 shows two probability curves having the same mean but different standard deviations.

Other useful terms that are necessary to have an introductory ability to describe a set of variables are the median, mode, and variance. The **median** is the value in a set of data that separates the lower half from the upper half. The **mode** is the value that is repeated most often in the set; if two values share this distinction, the data are said to be **bimodal.** The **variance,** like the standard deviation, measures the scatter of data about the mean; its actual mathematical formula is simply the square of the standard deviation.

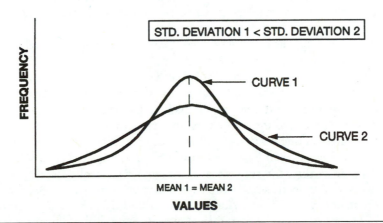

Figure A–2   Comparison of Two Different Probability Functions with the Same Mean.

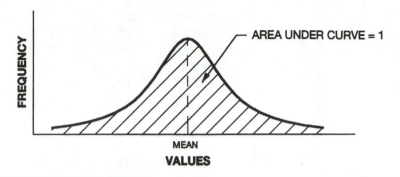

Figure A–3   Normal Probability Distribution.

One of the most common probability distributions in statistics is the **normal distribution curve,** or the normal curve. Students may also be familiar with this distribution (regarding the grades in their classes) by its nickname—the "bell" curve. The graph of this curve is the most important in the field of probability because it has been found to accurately represent the frequency of occurrence in almost all random variables. The normal distribution, or normal curve, has the following features (see Figure A–3):

- It is symmetrical about the y-axis, with the mean, mode, and median occurring at its peak.
- It asymptotically approaches the x-axis in each direction.
- The area under the curve equals one.

The area under these normal curves is an extremely important feature because it can be related to the probability of an event's occurring at a particular value of x. Statisticians have developed a correlation of area under the normal curve to the number of standard deviations (+ or −) that a particular value is away from the mean. For instance, the area encompassed under any normal curve from −1σ to +1σ is known to be 68.3% of the total area. The area encompassed under the normal curve from −2σ to +2σ is known to be 95.5% of that same area.

So far we have talked about probability in a generic, nonstructured approach. But if we think of the failure mechanism in structures, we can begin to see that it depends on two different factors: the strength, or **resistance,** of a member and the **loads** acting on that member. From a statistical perspective, the two factors can be viewed as independent random variables, neither of which is time dependent. In reality, the resistance of a member generally decreases with time while loads tend to increase.[2] *Failure can be simplified by being viewed as occurring when the structural resistance of a member is exceeded by the load effects on that member.* In Figure A–4 we see two normal curves representing the structural resistance, R, and load effects, Q. It must be realized that in the realm of probability we must believe that a chance exists for a member's resistance to occur anywhere along the R

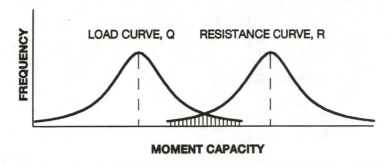

Figure A–4   Structural Probability Curves. Loading Distribution and Member Resistance Curves.

curve and that a load effect can likewise occur anywhere along the Q curve. When an event occurs whereby $Q > R$, there is no margin of safety and the member can be considered to have failed. The probability of failure can be viewed as a function of the area under the load (Q) and resistance (R) curves. Realizing that normal curves are asymptotic to the x-axis, the student should be aware that the curves *always* intersect. This essentially means that no matter how strong our member is, there is some possibility (however infinitely small) that failure might occur.

A.4   Limit States

When a member no longer functions as it was designed, it is said to have exceeded a limit state. In LRFD design (as in ASD), two types of limit states are discussed: strength limit states and serviceability limit states. Designers are primarily concerned with strength limit states such as load-carrying capability (which may specifically refer to moment capacity, shear capacity, or plastic hinge formation). Serviceability limit states refer primarily to deflection, vibration, and drift. The focus of most codes is on the strength limit states because of the great concern for public safety. Serviceability requirements tend to be much broader in scope and depend greatly on the designer's judgment.

In LRFD, the general strength limit state that should not be exceeded is referred to as the safety margin, Z. If the following equation accurately describes structural capacity,

$$Z = R - Q$$

where

R = resistance function

Q = load function

then a limit state is violated when $Z < 0$.

Limit state violations in Allowable Stress Design (ASD) were expressed in

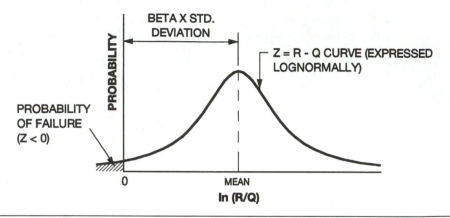

Figure A–5   Illustration of the Reliability Index or the "Margin of Safety."

terms of a factor of safety against a particular failure. From a design perspective, the major shortcoming with the ASD philosophy is that the factors of safety for various behaviors (bending, shear, etc.) all have different values. These values emerged from years of practical design experience, not from a rational basis. This has led to designs in which members experiencing different conditions (such as different loading types) would actually be very different in terms of a safety margin against failure. This is not to say that the ASD method has worked poorly; on the contrary, it has worked well. But the philosophy, while working well in terms of public safety, is providing nonuniform reliability to all members.

In the LRFD method, a measure of structural integrity called the reliability index, ß, is used in place of the traditional factor of safety. This reliability index is essentially the number of standard deviations from a combined frequency-density curve of resistance minus loading (with both functions being expressed lognormally on the same axis). A typical strength limit state violation can then be regarded as a probability of the safety margin, Z, being less than zero. This combined normal curve can be seen in Figure A–5.

The reliability index, ß, is essentially the number of standard deviations that the mean value of $\ln(R/Q)$ (or the expected value of this mean) lies away from the point of failure (where $Z < 0$). The higher the reliability index becomes, the wider is the spread between failure and the mean value of $R - Q$ [or $\ln(R/Q)$]. One of the main goals of the LRFD philosophy is the establishment of uniform reliability (or safety margin) between all types of members subjected to all types of loads.

A.5   Load and Resistance Factor Design (LRFD) Philosophy

LRFD is a unique philosophy because it establishes a uniform reliability for all members in a structure based on a probabilistic approach to load factors and resistances. Pioneered by Galambos,[3,4] the LRFD method strives to achieve the proper emphasis on both loads and different structural behaviors so as to design more cost-effective structures.

The design method for LRFD boils down to one general equation:

Design strength ≥ Required resistance

or

$$\phi R_n \geq \sum \delta_i Q_i$$

where

ϕ = resistance factor ($\phi < 1$)

R_n = nominal resistance of member

δ_i = Load factors ($\phi_i > 1$)

Q_i = different load effects

The required resistance is computed using the highest of several different load combinations as set forth in Section A4 of the LRFD specification. These load combinations will be discussed further in the next section. The required resistance may be the factored moment, shear, or axial load applied to a particular member.

The load factors used for the required resistance by the AISC specification were developed by the American National Standards Institute[5] (ANSI A58.1, now ASCE 7–88) for the purpose of specifying minimum acceptable load criteria for buildings. The load factors for individual load types (dead, live, wind) stem from the previously mentioned reliability index. Target levels for ß range from 3.0 for gravity loads to 1.75 for combinations with earthquake loading.

The nominal resistance of a member is the calculated strength based on standard techniques. It should be realized that now we utilize full material properties (such as F_y) when calculating member capacities.

The resistance factors are developed based on the variability of each resistance, with the goal of establishing a uniform reliability between members. For example, the ϕ factors will be lower for columns than for beams because column behavior is more sensitive to factors such as initial crookedness and therefore will be given a lower resistance factor. These resistance factors will be unique to each type of behavior for which we are designing.

A.6 Load Combinations

Much work[6] in probabilistic load criteria has culminated in the publication of the ANSI Standard as we referenced previously. LRFD uses the basic load combinations as suggested by this standard as the minimum requirements for load-carrying structures. We have already mentioned that LRFD states that the design strength of a member has to be greater than or equal to the required resistance. The required resistance is the effect (moment, shear, etc.) caused by the highest factored load combination as stated in the ANSI publication or in Section A4.1 of the AISC specification. These load combinations, with their respective load factors, are as follows:

$1.4D$	AISC LRFD (A4–1)
$1.2D + 1.6L + 0.5(L_r \text{ or } S \text{ or } R)$	AISC LRFD (A4–2)
$1.2D + 1.6(L_r \text{ or } S \text{ or } R) + (0.5L \text{ or } 0.8W)$	AISC LRFD (A4–3)
$1.2D + 1.3W + 0.5L + 0.5(L_r \text{ or } S \text{ or } R)$	AISC LRFD (A4–4)
$1.2D + 1.5E + (0.5L \text{ or } 0.2S)$	AISC LRFD (A4–5)
$0.9D - (1.3W \text{ or } 1.5E)$	AISC LRFD (A4–6)

with

D = gravity dead loads

L = gravity live loads

L_r = roof live loads

W = wind load

E = earthquake load

S = snow load

R = load to initial rainwater or ice

A load combination may have up to three separate parts, consisting of the factored dead load effect, the 50-year maximum live load effect, and possibly another live load effect labeled as an arbitrary-point-in-time (APT) value. Such APT values are typically only a fraction of their design values, since this would reflect a small probability that they would occur simultaneously with the other 50-year maximum live loads or full gravity loads.

A.7   AISC Column Design Philosophy Using LRFD

The AISC formulas for column design are the result of years of research committed to finding a formula that accurately predicts column behavior in all slenderness ranges. As discussed in Chapter 5, the AISC philosophy recognizes that the behavior of a column changes from elastic buckling to inelastic buckling as the slenderness decreases. Because of this change in column behavior, the AISC sets a value of a slenderness parameter (termed Ω_c) as the dividing line between the two aforementioned ranges. The value of this slenderness parameter is actually a ratio of yield stress to Euler buckling stress and can be written as

$$\Omega_c^2 = F_y/F_{\text{euler}} \quad \text{where} \quad F_{\text{euler}} = \pi^2 E/(kl/r)^2$$

or

$$\Omega_c = (KL/r\pi)\sqrt{F_y/E} \quad \text{(AISC LRFD Eq. 2–4)}$$

The formulas that the LRFD specification designates to handle column behavior are similar in nature to those that we have discussed previously in Chapter 5. The formulas calculate the critical column stress, F_{cr}, over slenderness ranges in the elastic and inelastic regions.

For slender columns (when $\Omega_c > 1.5$), the elastic behavior mode controls, and the following equation is used to calculate the critical stress:

$$F_{cr} = (.877/\Omega_c^2)F_y \quad \text{(AISC LRFD Eq. E2–3)}$$

The .877 modifier in this equation is used to account for initial out-of-straightness. This modifier stems from the safety factor (23/12) used in the denominator of AISC Eq. E2–2 of the ASD specification. The nominal safety factor in the allowable stress method was 1.67, but this increased to 1.92 due to column crookedness. To remain compatible with ASD, the LRFD method multiplies the elastic formula by .877, which is approximately the modifier derived from dividing the nominal column safety factor (1.67) by the long column safety factor (1.92).

For shorter (and less slender) columns (when $\Omega_c \leq 1.5$), inelastic column behavior controls and the LRFD specification presents the following equation:

$$F_{cr} = (.658\Omega_c^2)F_y \quad \text{(AISC LRFD Eq. E2–2)}$$

The reader should notice that the aforementioned AISC equations E2–2 and E2–3 become equal when the slenderness parameter, Ω_c, equals 1.5. At this point both equations become equal to $.39F_y$, which accounts for the effects of residual stress. This AISC philosophy is outlined in Figure A–6.

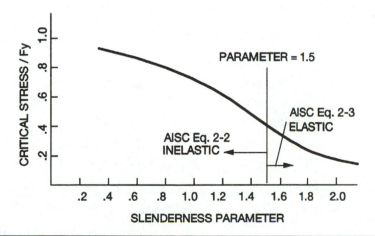

Figure A–6   LRFD Column Philosophy.

A.8   Column Design per LRFD

The LRFD method for column design follows the same format as was discussed in the previous sections, which is stated as follows:

Design strength ≥ required resistance

For columns, this general format will take the following specific form:

$$\phi_c P_n \geq P_u$$

where

ϕ_c = .85, column reduction factor

$P_n = A_g F_{cr}$, nominal column strength

P_u = maximum factored column load

As we have stated before, there are two general types of problems encountered in steel design: evaluation and design. In the evaluation problem, you will be given a member and corresponding load and asked to determine its adequacy. For the design problem, you will be given loads and asked to design the member size.

The evaluation problem is very easy to perform, as attested to by the following examples. A comparison of a column's design strength ($\phi_c P_n$) to the factored load (P_u) applied to that column will prove whether the member is adequate. The following examples will provide some insight to these problems.

EXAMPLE A.1

Calculate the maximum factored load (P_u) that a W 14 × 53 column can hold if it is 20 ft long with a $K = 1.0$. The steel is A36.

First, consider the section properties:

A W 14 × 53 has $A = 15.6$ in.^{2}

$r_x = 5.89$ in. and $r_y = 1.92$ in.

r_y will control since column is not braced

$Kl/r_y = 1.0 \times 20$ in. x 12 in./ft/1.92 in. = 125

$\Omega_c = (125/\pi)\sqrt{36{,}000 \text{ psi}/29 \times 10^6 \text{ psi}} = 1.401$

$\Omega_c = 1.401 \leq 1.5$ used LRFD Eq. (E2–2)

$F_{cr} = [.658^{(1.401)^2}]36$ ksi = 15.81 ksi

Max $P_u = .85 \times 15.81$ ksi $\times 15.6$ in.^{2} = 209.6 *kips*

Note: When bracing is used, the slenderness parameters, Ω_c, about both the strong and weak axes should be checked with the larger controlling the design.

EXAMPLE A.2

(a) Calculate the maximum factored load, P_u, for a W 24 × 104 that is 20 ft long with $K = .80$. (b) Then, recalculate if the column is braced at midheight in the weak axis (still assume $K = 0.80$). Use A36 steel.

a.   First, consider the section properties:

For a W 24 × 104, $A = 30.6$ in.^{2}

$r_x = 10.1$ in.

$r_y = 2.91$ in.

Kl/r_y controls = .8(20 ft × 12 in./ft)/2.91 in. = 65.98

$\Omega_c = (Kl/r\pi)\sqrt{F_y/E}$

$\Omega_c = (65.98/\pi)\sqrt{.00124} = .7399 < 1.5$

$F_{cr} = [.658^{(.7399)^2}]36$ ksi = 28.63 ksi

Max $P_u = .85 \times 28.63$ ksi × 30.6 in.^{2} = 744.6 *kips*

b.   Braced midheight in weak direction:

$Kl/r_x = .8(20$ ft × 12 in./ft)/10.1 in. = 19.01

$Kl/r_y = .8(10$ ft × 12 in./ft)/2.91 in. = 32.99 still controls

$\Omega_c = (32.99/\pi)\sqrt{.00124} = .3697 < 1.5$

$F_{cr} = [.658^{(.3697)^2}]36$ ksi = 34 ksi

Max $P_u = .85 \times 34$ ksi × 30.6 in.^{2} = 884.4 *kips*

The design problems are, as usual, a little more difficult than when evaluating a member. This appendix will proceed as though students will not have the many available design tables at their disposal. The proposed design procedure is a trial-and-error approach, but will be found to be very quick using LRFD. If the reader realizes that the critical stress, F_{cr}, is dependent only on the slenderness parameter, and that the only variable in the equation for Ω_c (given a particular steel) is the radius of gyration, the selection of where to begin the design process starts with an assumption of the radius of gyration, r. Although the value of r might be subject to a large variation over a particular section, we will find the calculation of F_{cr} using AISC Eqs. (E2–2) and (E2–3) to be very fast. The step-by-step process for this design procedure is as follows:

1.   Select an r value general to the majority of members under consideration.
2.   Using $r_{assumed}$, calculate Ω_c.
3.   Calculate F_{cr}, using either (E2–2) or (E2–3).
4.   Calculate $A_{g\ assumed} = P_u/\phi_c F_{cr}$.

5. Choose trial member based on $A_{g \text{ assumed}}$.
6. Calculate F_{cr} of trial member and compare $\phi_c F_{cr} A_g$ to P_u.
7. If $\phi_c F_{cr} A_g \ggg P_u$, choose the smaller; if $\phi_c F_{cr} A_g < P_u$, choose the larger.

The following example illustrates the use of this procedure in a column design problem.

EXAMPLE A.3

Find the most economical W12 section to hold a compressive service load of 120 kips. Steel is A36, $K = 1.0$, $l = 16$ ft. Assume that the live load to dead ratio equals 1.0 (combination A4–2 controls) and that the column is unbraced.

Looking at W12 sections, assume a general r_y value of 3.0.

$Kl/r_y = (1.0)(16 \text{ ft} \times 12 \text{ in./ft})/3.0 \text{ in.} = 64$

$\Omega_c = (64/\pi)\sqrt{36,000/29 \times 10^6} \text{ psi} = .7177 < 1.5$ (AISC Eq. E2–2)

$F_{cr} = [.658^{(.7177)^2}]36 \text{ ksi} = 29.02 \text{ ksi}$

Calculate the factored load, P_u, based on equal live and dead loads:

$P_u = 1.6(60) + 1.2(60) = 168 \text{ kips}$

$A_{g \text{ assumed}} = 168 \text{ kips}/(.85)(29.02 \text{ ksi}) = 6.81 \text{ in.}^2$

W 12 × 26 suffices, but $r_y = 1.51$, about 1/2 of our assumption—probably no good.

Check $Kl/r_y = (1.0)(16 \text{ ft} \times 12 \text{ in./ft})/1.51 \text{ in.} = 127.15$. Therefore, $\Omega_c = 1.426$.

$F_{cr} = [.658^{(1.426)^2}]36 \text{ ksi} = 15.37 \text{ ksi}$

$\phi_c A_y F_{cr} = (.85)(7.65)(15.37) = 99.9 \text{ kips} < 168 \text{ kips}$ No good.

Try W 12 × 40 $A = 11.8 \text{ in.}^2$—$r_y = 1.93 \text{ in.}$ $Kl/r_y = 99.48$. Therefore, $\Omega_c = 1.116 < 1.5$:

$F_{cr} = [.658^{(1.116)^2}]36 \text{ ksi} = 21.38 \text{ ksi}$

$\phi_c F_{cr} A_g = .85(21.38 \text{ ksi})(11.8 \text{ in.}^2) = 214.4 \text{ kips} > 168 \text{ kips}$

At this point the reader might decide that a section smaller than a W 12 × 40 might work, since there seems to be a sizable gap between the required strength (168 kips) and the design strength (214.4 kips). Try a W 12 × 35— you'll find that this will not work and that the W 12 × 40 is therefore the most economical.)

In addition to this trial-and-error design procedure, the student should be aware that design tables are available that are very similar to those used in the ASD philosophy. These tables are found in Part 2 of the LRFD specification.

A.9   AISC Rolled-Beam Design Philosophy Using LRFD

The most basic requirement of beam design using the LRFD method is that design moment capacity, ϕM_n, must be greater than or equal to the required (factored) flexural strength, M_u. The required flexural strength may be referred to by some as the "ultimate moment." This most basic requirement can be simply stated as follows:

$$\phi M_n \geq M_u$$

where $\phi = .90$ for bending.

As discussed earlier in Chapter 6, bending strength is affected by unbraced length of the compression flange, the compactness of the member, and the axis of bending. Since rolled shapes are primarily bent about the strong axis and are compact, we will restrict our discussion of the LRFD requirements to those that focus on the unbraced length of the compression flange.

For the unbraced length of the compression flange, there are two lengths that dictate the limit state behavior of the section, L_p and L_r. The L_p limit is the dividing line delineating the unbraced length under which a beam can actually reach the plastic moment, M_p, and is given as follows:

$$L_p = 300 r_y / \sqrt{F_{yf}} \qquad \text{(AISC LRFD F1–4)}$$

where

r_y = radius of gyration

F_{yf} = flange yield strength of steel, ksi

The L_r limit is the unbraced length dividing elastic and inelastic buckling of the compression flange and is given in the AISC manual as follows:

$$L_r = \frac{r_y X_1}{F_{yw} - F_r} \sqrt{1 + \sqrt{1 + X_2 (F_{yw} - F_r)^2}} \qquad \text{(AISC LRFD F1–6)}$$

where

r_y, F_y are as previously defined

X_1, X_2 = beam buckling factors

F_r = compressive residual stress generally taken as 10 ksi

F_{yw} = yield strength of the web

The beam buckling factors, X_1 and X_2, are found with the section properties in the AISC tables or in the following equations, as given in Chapter F of the LRFD specification:

$$X_1 = \frac{\pi}{S_x} \sqrt{\frac{EGJA}{2}}$$

$$X_2 = 4 \frac{C_w}{I_y} \left(\frac{S_x^2}{GJ} \right)^2$$

Although these terms may be confusing to the average person, they can be thought of as simply terms describing the bending efficiency of a member.[7] As the X_1 term decreases the bending efficiency of the member increases, and as the X_2 term increases the efficiency increases.

For compact rolled beams, the AISC design philosophy for calculating nominal moment capacity, M_n, is shown in Figure A–7. This curve is described in the following list, where the three entries correspond to the three categories in the figure:

1. *Compact beams with unbraced length, $l_b < L_p$.* Beams in this region can achieve full plastic moment capacity; therefore, the nominal moment capacity in this region is the plastic moment, M_p.
2. *Compact beams with $L_p < l_b < L_r$.* Beams in this region exhibit a limit state of inelastic lateral torsional buckling, before being able to achieve M_p. The AISC gives the nominal moment capacity, Eq. (F1–3), as follows:

$$M_n = C_b \left[M_p - (M_p - M_r) \left(\frac{L_b - L_p}{L_r - L_p} \right) \right] \quad \text{(AISC LRFD F1–3)}$$

This equation is simply a linear interpolation between M_p and M_r, based on the unbraced length of beam. The moment capacity M_r requires some explanation at this point. M_r is the moment capacity of the beam as it reaches first yield, in this case due to lateral torsional buckling. As with columns, the consideration of residual stresses in the beam's compression flange is important, because it can affect the value of moment causing first yield. Thus M_r is expressed as

$$M_r = (F_y - F_r)S_x$$

where F_r = compressive residual stress in flange considered to be 10 ksi for rolled beams.

The student should be advised that residual stresses affect only the value of M_r, *not* the value of M_p, since plastic moment capacity considers yield over the entire area and residual stresses must be in equilibrium before loads are applied.

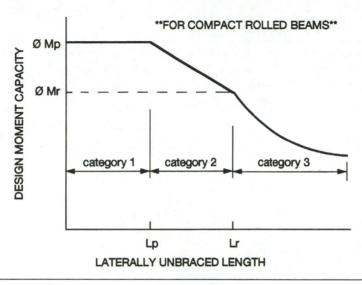

Figure A–7   Compact Beam Behavior Using LRFD Specifications.

3.   *Compact beams with $l_b > L_r$.*   This region contains beams that fail by exhibiting the behavior of elastic lateral torsional buckling. The beams in this region have long, slender unsupported lengths of the compression flange, and failure occurs before the sections reach yield. The nominal moment capacity of a beam in this region is equal to the critical buckling moment, M_{cr}. The AISC formula for this is

$$M_{cr} = C_b \frac{\pi}{l_b} \sqrt{\left(\frac{\pi E}{l_b}\right)^2 C_w I_y + E I_y G J} \qquad \text{(AISC LRFD F1–13)}$$

or

$$M_{cr} = \frac{C_b S_x X_1 \sqrt{2}}{l_b / r_y} \sqrt{1 + \frac{X_1^2 X_2}{2(l_b / r_y)^2}}$$

The following examples will demonstrate the procedure for calculating the adequacy of rolled-steel beams that fall into these three categories. Should a rolled section be noncompact (although as mentioned before this would be highly unlikely), the approach to calculating the nominal moment capacity, ϕM_n, is extremely similar to the aforementioned category 2. The nominal moment capacity will be calculated by choosing the most critical of the three potential modes of failure: lateral torsional buckling, flange buckling, and web buckling.

EXAMPLE A.4

Determine the adequacy of the W 12 × 87 shown in the accompanying figure if the compression flange is braced only at the supports. The load shown is 60% live and 40% dead and unfactored. The steel is A36. Assume $C_b = 1.0$, since the beam is simply supported.

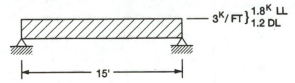

$D = 0.40\ (3\text{ kips/ft}) = 1.2\text{ kips/ft}$

$L = 0.60\ (3\text{ kips/ft}) = 1.8\text{ kips/ft}$

Load combination A4–2 controls; therefore, the factored load is

$1.2(1.2\text{ kips/ft}) + 1.6(1.8\text{ kips/ft}) = 4.32\text{ kips/ft}$

$M_u = 4.32\text{ kips/ft}(15\text{ ft})^2/8 = 121.5\text{ kip-ft}$

Check to see whether a W 12 × 87 is compact:

For the flanges: $b_f/2t_f = 7.5$, which is less than $65/\sqrt{F_y}$

(therefore, flange is compact).

For the web: $h_c'/t_w = 18.9$, which is less than $640/\sqrt{F_y}$
(therefore, web is compact).

Unbraced length, $l_b = 15$ ft.

$L_p = 300 r_y/\sqrt{F_y}.$

$L_p = 300(3.07\text{ in.})/\sqrt{36\text{ ksi}} = 153.5\text{ in.}\quad\text{or}\quad 12.8\text{ ft.}$

$L_r = \left[(r_y X_1)/(F_y - F_r)\right]\sqrt{1+\sqrt{(1+x_2(F_y - F_r))}}.$

$L_r = 3.07\text{ in}(3880\text{ ksi})/26\text{ ksi}\sqrt{1+\sqrt{1+.000586(36\text{ ksi}-10\text{ ksi})^2}}$
    $= 676.6\text{ in.}\quad\text{or}\quad 56.4\text{ ft.}$

Therefore, $L_p < l_b < L_r$, so use category 2 (AISC LRFD Eq .F1–3):

$M_p = F_y Z_x$

$M_p = 36\text{ ksi}(132\text{ in.}^3)/12\text{ in./ft} = 396\text{ kip-ft}$

$$M_r = (F_y - F_r)S_x$$

$$M_r = (36 \text{ ksi} - 10 \text{ ksi})(118 \text{ in.}^3)/12 \text{ in./ft} = 255.7 \text{ kip-ft}$$

$$M_n = 1.0[396 \text{ kip-ft} - (396 \text{ kip-ft} - 255.7 \text{ kip-ft})(15 \text{ ft} - 12.8 \text{ ft})/$$
$$(56.4 \text{ ft} - 12.8 \text{ ft})] = 388.9 \text{ kip-ft} \qquad \text{(AISC Eq. F1–3)}$$

$$\phi M_n = .90(388.9 \text{ kip-ft}) = 350 \text{ kip-ft} > 121.5 \text{ kip-ft} \qquad \text{OK.}$$

EXAMPLE A.5

Recalculate the design moment capacity of the beam in Example A.4, if the W 12 × 87 has its compression flange braced at midspan.

In this case the unbraced length, l_b, is reduced to 7.5 ft, which is less than the L_p value of 12.8 ft calculated earlier. Therefore, this case falls into category 1 (which was a compact beam with $l_b < L_p$), and the nominal moment capacity is the plastic capacity.

$$M_n = M_p = Z_x F_y$$

$$M_n = 132 \text{ in.}^3 \times 36 \text{ ksi} = 4752 \text{ kip-in.} \quad \text{or} \quad 396 \text{ kip-ft}$$

$$\phi M_n = .90 \times 396 \text{ kip-ft} = 356.4 \text{ kip-ft}$$

EXAMPLE A.6

Calculate the adequacy of a W 10 × 33 that is 25 ft long and subject to a factored load of 1.2 kips per foot. The beam is unbraced and made from A242 steel ($F_y = 50$ ksi). Assume $C_b = 1.0$.

In this problem we can first check to see whether the beam is compact. Checking the flanges, we see that $b_f/2t_f = 9.1$, which is just less than $65/\sqrt{F_y}$ (=9.2), so the flange is compact. Checking the web, we see that $h_c/t_w = 27.1$, which again is less than $640/\sqrt{F_y}$; therefore, the web is also compact. Next, let's check the unbraced length criteria:

$$l_b = 25 \text{ ft}$$

$$L_p = 300(1.94 \text{ in})/\sqrt{50} = 82.3 \text{ in} \quad \text{or} \quad 6.86 \text{ ft}$$

$$L_r = \left[(r_y X_1)/(F_y - F_r)\right]\sqrt{1 + \sqrt{1 + x_2(F_y - F_r)^2}}$$

$$L_r = 19.7 \text{ ft}$$

Therefore, since $l_b > L_r$, elastic lateral torsional buckling will predominate and the nominal moment is calculated as the critical moment, M_{cr}, which is AISC Eq. (F1–13) as follows:

To reduce the equation's length, the quantity l_b/r_y is precalculated as 25 × 12 in./ft − 1.94 = 154.6

$$M_{cr} = \frac{1.0\left(35 \text{ in.}^3\right)\left(2710 \text{ ksi}\right)\sqrt{2}}{154.6}\sqrt{1 + \frac{(2710 \text{ ksi})^2\,(.00251)}{2(154.6)^2}} = 85./\text{kip-ft}$$

Therefore ϕM_n = .90 × 85.1 kip-ft = 76.6 kip-ft
Since $M_u = 1.2 \times (25 \text{ ft})^2/8 = 93.75$ kip-ft > 76.6 kip-ft   The section is no good.

A.10   AISC Rolled-Beam Design Using LRFD

As with ASD, numerous beam design tables and charts are readily available to the experienced designer for use in LRFD. However, in this brief overview of the LRFD method, the author will assume that the reader does not have access to such charts and therefore will approach beam design using the principles as developed in the earlier section.

The following is a basic format to help the beginner in LRFD of rolled beams. As mentioned earlier in this chapter, the primary concern of beam design is to satisfy the following criterion:

$$M_u \le \phi M_n \qquad \text{or} \qquad M_u/\phi \le M_n$$

In this case, M_u and ϕ are known (or can be found) and the chore becomes finding a beam that equals or exceeds M_n.

From Section A.9, we know that compact rolled beams fall into one of three categories, based on the unbraced length of the compression flange, l_b. Since l_b will be given or assumed in any problem, the task begins by looking at the dividing lines between plastic moment capacity and inelastic buckling, L_p, and inelastic and elastic buckling, L_r. The first look will be to see whether wide-flange sections are even close to being in category 1 (in which beams can reach their full plastic moment), where $l_b < L_p$. Using this strategy, we can set $l_b = L_p$ and solve for a minimum value of r_y as follows:

$$r_{y\,min} = l_b \times \sqrt{F_y/300}$$

If a large majority of our sections meet the minimum requirement for r_y, they stand a very good chance of being in category 1 and therefore

$$M_n = M_p = Z_x F_y$$

Solving for Z_x yields

$Z_{x \text{ req'd}} = M_n/F_y$

The following example will demonstrate this design technique for rolled-steel beams.

EXAMPLE A.7

Design a W12 section to hold a factored moment of 400 kip-ft. The steel is A36, and the beam is 12 feet long and unbraced ($C_b = 1.0$).

$l_b = 12 \text{ ft} = 144 \text{ in.}$

$r_{y \text{ min}} = 144 \times \sqrt{36}/300 = 2.88$

From the W12 charts, the majority of W12's have r_y values in excess of the 2.88 minimum value. If our sections have an r_y above this value, the following design is true:

$M_u = 400 \text{ kip-ft}$

$M_n = 400 \text{ kip-ft}/.90 = 444.4 \text{ kip-ft}$

$M_n = Z_x F_y$ (if category 1)

$Z_{x \text{ req'd}} = (444.4 \text{ kip-ft} \times 12 \text{ in./ft})/36 \text{ ksi} = 148.1 \text{ in.}^3$

From the table, select W 12×106 ($Z_x = 164 \text{ in.}^3$). Since a W 12×106 has an $r_y = 3.11$, it does fall in category 1, since $l_b < L_p$.

$L_p = 300(3.11)/\sqrt{36} = 155.5/12 = 12.9 \text{ ft}$

Therefore, $M_n = 164 \text{ in.}^3 \times 36 \text{ ksi} = 5904 \text{ kip-in.}$ or 492 kip-ft and

$\phi M_n = .90 \times 492 = 442.8 \text{ kip-ft} > 400 \text{ kip-ft}$ It works.

Example A.7 was nice, in that all of our assumptions were correct, but what would have happened if our beam had not fallen in category 1? The simple answer is that our design problem would have become more difficult. If we remember that the nominal moment capacity decreases as $l_b > L_p$, however, we can use logic to help us in our assumptions when our beam falls into categories 2 and 3, as outlined in Section A.9.

A beam is designed according to category 2 if $L_p < l_b < L_r$ and the nominal moment in category 2 ranges from M_p (when the unbraced length is equal to L_p) to M_r (when the unbraced length is equal to L_r).

Therefore, if the design fails to meet category 1 requirements, gauge the upper and lower limits of category 2 by solving the following:

$Z_{x \text{ req'd}} = M_n/F_y$ (upper limit category 2)

$S_{x \text{ req'd}} = M_n/(F_y - F_r)$ (lower limit category 2)

Next, realizing that your first assumption of a beam's falling in category 1 left you with a section's value of L_p and L_r, try to gauge where your section might fall. If your l_b is barely over L_p, then solve for a section based on its $Z_{x \text{ req'd}}$. If your **unbraced length,** l_b, is barely greater than L_r, then solve for a section based on its $S_{x \text{ req'd}}$. Example A.8 will illustrate this technique.

EXAMPLE A.8

Design a W12 section to hold a factored moment of 220 kip-ft. The steel is A36, and the beam is 12 ft long and unbraced ($C_b = 1.0$).

$l_b = 144$ in.

$r_{y \min} = 144 \times \sqrt{36}/300 = 2.88$ in.

$M_u = 220$ kip-ft

$M_n = 220$ kip-ft/0.90 = 244.4 kip-ft

(Assume category 1) $M_n = Z_x F_y$

$Z_{x \text{ req'd}} = (244.4 \text{ kip-ft} \times 12 \text{ in./ft})/36 \text{ ksi} = 81.5 \text{ in.}^3$

From tables choose W 12×58 ($Z_x = 86.4$ in.^{3}). But W 12×58 has $r_y = 2.51$ in., which is less than $r_{y \min}$; therefore, this section does not fall in category 1.

W 12×58 $L_p = 300(2.51 \text{ in.})/36$ ft = 125.5 in. = 10.4 ft

$L_r = 38.4$ ft

10.4 ft < l_b = 12 ft < 38.4 ft

So a W 12×58 falls into category 2, but just barely. Let's try this section to see whether it does work anyway, because the plastic section modulus is somewhat larger than needed.

$M_n = C_b \{M_p - (M_p - M_r)[(L_b - L_p)/(L_r - L_p)]\}$ (AISC LRFD F1–3)

$C_b = 1.0$

$M_p = Z_x F_y = 86.4$ in.$^3 \times 36$ ksi/12 in./ft = 259.2 kip-ft

$M_r = (36 \text{ ksi} - 10 \text{ ksi}) 78 \text{ in.}^3/(12 \text{ in./ft}) = 169$ kip-ft

$L_b = 12$ ft

$$L_p = 10.4 \text{ ft}$$

$$L_r = 38.4 \text{ ft}$$

$$M_n = 1.0\left\{259.2 \text{ kip-ft} - \left(259.2 \text{ kip-ft} - 169 \text{ kip-ft}\right)\left(\frac{12 \text{ ft} - 10.4 \text{ ft}}{38.4 \text{ ft} - 10.4 \text{ ft}}\right)\right\}$$

$$= 254 \text{ kip-ft}$$

Since $M_n = 254$ kip-ft > 244.4 kip-ft, W 12 × 58 works. (It is also compact.)

(If the W 12 × 58 had failed, we would logically have tried a larger section; and if a W 12 × 58 was greatly overdesigned, we would have scaled down.)

A.11   AISC Tension Member Philosophy Using LRFD

Tension member design using the LRFD method again follows the same general principles that we have discussed in Chapter 4. The two mechanisms of strength-based limit states are failure by yielding over the gross area (A_g) and failure by fracture through the effective net area (A_e). Both locations must be checked to determine which is most critical. The basic requirement behind LRFD in tension members is exactly the same as previously mentioned: Make sure your design strength is greater than your required strength. This requirement for tension members can be expressed as

$$P_u \le \phi_t P_n$$

where

P_u = maximum load from factored load combinations

ϕ_t = reduction factor (as follows) = .90 across the gross area; = .75 across the net area

P_n = nominal (calculated) tension capacity using full material properties

This can be further broken down into specific equations across the gross and effective net area:

Gross area: $P_u \le \phi_t A_g F_y$

where

A_g = gross area, in.^{2}

F_y = yield stress, ksi or psi

ϕ_t = .90

Effective net area: $P_u \le \phi_t A_e F_u$

where

$A_e = A_{net} \times$ reduction factor, U

$F_u = $ ultimate tensile

$\phi_t = .75$

The following example will demonstrate the use of the LRFD method in a typical evaluation problem.

EXAMPLE A.9

The tension member shown in the accompanying figure is subjected to a *DL* of 21.25 kips and a *LL* = 63.75 kips (both service). Calculate the member's adequacy across the gross and net sections. The steel is A36, and the bolts are 3/4 in. in diameter.

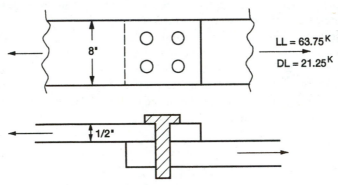

$A_{gross} = 8$ in. $\times$ 1/2 in. $= 4$ in.^{2}

$A_{net} = 4$ in.$^2 - 2(3/4$ in. $+ 1/8$ in.$)(1/2$ in.$) = 3.125$ in.^{2}

$U = 1.0$ since member a flat plate

$A_e = 1.0 \times 3.125$ in.$^2 = 3.125$ in.^{2}

Design strength over gross area $= .9(4$ in.$^2)(36$ ksi$) = 129.6$ kips (controls)

Design strength over effective net area $= .75(3.125$ in.$^2)(58$ ksi$) = 135.9$ kips

Calculating of factored load combinations:

1.4 (21.25 kips) = 29.75 kips

1.2 (21.25 kips) + 1.6(63.75 kips) = 127.5 kips (controls)

AISC Eqs. (A4–3) through (A4–6) obviously don't control:

$P_u = 127.5$ kips < 129.6 kips   Section is adequate.

A.12   AISC Tension Member Design Using LRFD

The design of tension members in LRFD incorporates the use of the direct stress equation across either of the following two equations:

$$A_g = P_u/(\phi_t F_y) \quad \text{for gross area}$$

or

$$A_e = P_u/(\phi_t F_u) \quad \text{for effective net area}$$

Design is carried out in the same manner as was previously discussed in Chapter 4, checking across both the effective net area and the gross area.

EXAMPLE A.10

Design the thickness of the flat plate tension member shown in the figure, if the load is 55 kips. Assume an *L/D* ratio of 1.0. The bolts are 3/4 in. in diameter, and the steel is A36.

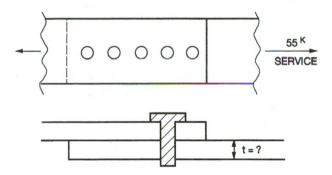

Calculate maximum factored loads, realizing that live load = dead load (since *L/D* = 1.0). Therefore, *LL* = 27.5 kips, *DL* = 27.5 kips.

By inspection, load combination (A4–2) controls:

$$1.2(27.5 \text{ kips}) + 1.6(27.5 \text{ kips}) = 77 \text{ kips} = P_u$$

Gross Area

$$A_{\min} = 77 \text{ kips}/[(.90)(36 \text{ ksi})] = 2.376 \text{ in.}^2$$

$$A_g = 6 \times t = 2.376 \text{ in.}^2, \quad \text{therefore, } t = .396 \text{ in.}$$

Effective Net Area

$$U = 1.0 \text{ (flat plate)}$$

$$A_{\min} = 77 \text{ kips}/[(.75)(58 \text{ ksi})] = 1.77 \text{ in.}^2$$

$$A_e = 6t - .875t = 5.125t = 1.77 \text{ in.}^2, \qquad t = .345 \text{ in.}$$

Gross area thickness controls (.396 in.), since it is larger; therefore, the practical choice would be a 7/16-in. thick plate.

REFERENCES

1. Aine Brazil, Robert DeScenza, and Thomas Scarangello, "LRFD: Still Waiting," *Civil Engineering,* July 1991, pp .46–48.
2. R.E. Melchers, *Structural Reliability Analysis and Prediction* (New York: Ellis Horwood Limited Publishers, 1987), p.32.
3. Theodore Galambos and M.K. Ravindra, "Proposed Criteria for Load and Resistance Factor Design," *Engineering Journal,* AISC (1981), vol. 18, no. 3: 74–82.
4. M.K. Ravindra and Theodore Galambos, "Load and Resistance Factor Design for Steel," *Journal of the Structural Division,* ASCE, September 1978, pp. 1337–1353.
5. ASCE, *American Society of Civil Engineers Minimum Design Loads for Buildings and Other Structures,* ASCE 7–88 (formerly ANSI A58.1), ASCE, New York, 1990.
6. ASCE, *American Society of Civil Engineers Minimum Design Loads for Buildings and Other Structures,* ASCE 7–88 (formerly ANSI A58.1), ASCE, New York, 1990.
7. Peter Hoadley, "Practical Significance of LRFD Beam Buckling Factors," *Journal of Structural Engineering,* ASCE, March 1991, pp. 988–995.

B

SELECTED STEEL SECTION TABLES

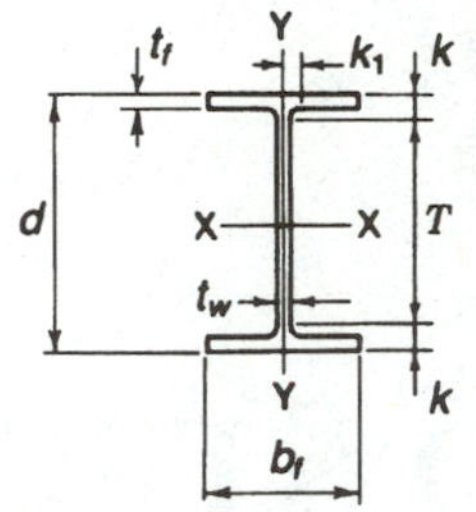

W SHAPES
Dimensions

Designation	Area A	Depth d		Web Thickness t_w		Web $\frac{t_w}{2}$	Flange Width b_f		Flange Thickness t_f		Distance T	Distance k	Distance k_1
	In.^{2}	In.		In.		In.	In.		In.		In.	In.	In.
W 16× 31	9.12	15.88	15 7/8	0.275	1/4	1/8	5.525	5 1/2	0.440	7/16	13 5/8	1 1/8	3/4
× 26	7.68	15.69	15 3/4	0.250	1/4	1/8	5.500	5 1/2	0.345	3/8	13 5/8	1 1/16	3/4
W 14×730[a]	215.0	22.42	22 3/8	3.070	3 1/16	1 9/16	17.890	17 7/8	4.910	4 15/16	11 1/4	5 9/16	2 3/16
×665[a]	196.0	21.64	21 5/8	2.830	2 13/16	1 7/16	17.650	17 5/8	4.520	4 1/2	11 1/4	5 3/16	2 1/16
×605[a]	178.0	20.92	20 7/8	2.595	2 5/8	1 5/16	17.415	17 3/8	4.160	4 3/16	11 1/4	4 13/16	1 15/16
×550[a]	162.0	20.24	20 1/4	2.380	2 3/8	1 3/16	17.200	17 1/4	3.820	3 13/16	11 1/4	4 1/2	1 13/16
×500[a]	147.0	19.60	19 5/8	2.190	2 3/16	1 1/8	17.010	17	3.500	3 1/2	11 1/4	4 3/16	1 3/4
×455[a]	134.0	19.02	19	2.015	2	1	16.835	16 7/8	3.210	3 3/16	11 1/4	3 7/8	1 5/8
W 14×426[a]	125.0	18.67	18 5/8	1.875	1 7/8	15/16	16.695	16 3/4	3.035	3 1/16	11 1/4	3 11/16	1 9/16
×398[a]	117.0	18.29	18 1/4	1.770	1 3/4	7/8	16.590	16 5/8	2.845	2 7/8	11 1/4	3 1/2	1 1/2
×370[a]	109.0	17.92	17 7/8	1.655	1 5/8	13/16	16.475	16 1/2	2.660	2 11/16	11 1/4	3 5/16	1 7/16
×342[a]	101.0	17.54	17 1/2	1.540	1 9/16	13/16	16.360	16 3/8	2.470	2 1/2	11 1/4	3 1/8	1 3/8
×311[a]	91.4	17.12	17 1/8	1.410	1 7/16	3/4	16.230	16 1/4	2.260	2 1/4	11 1/4	2 15/16	1 5/16
×283[a]	83.3	16.74	16 3/4	1.290	1 5/16	11/16	16.110	16 1/8	2.070	2 1/16	11 1/4	2 3/4	1 1/4
×257[a]	75.6	16.38	16 3/8	1.175	1 3/16	5/8	15.995	16	1.890	1 7/8	11 1/4	2 9/16	1 3/16
×233[a]	68.5	16.04	16	1.070	1 1/16	9/16	15.890	15 7/8	1.720	1 3/4	11 1/4	2 3/8	1 3/16
×211	62.0	15.72	15 3/4	0.980	1	1/2	15.800	15 3/4	1.560	1 9/16	11 1/4	2 1/4	1 1/8
×193	56.8	15.48	15 1/2	0.890	7/8	7/16	15.710	15 3/4	1.440	1 7/16	11 1/4	2 1/8	1 1/16
×176	51.8	15.22	15 1/4	0.830	13/16	7/16	15.650	15 5/8	1.310	1 5/16	11 1/4	2	1 1/16
×159	46.7	14.98	15	0.745	3/4	3/8	15.565	15 5/8	1.190	1 3/16	11 1/4	1 7/8	1
×145	42.7	14.78	14 3/4	0.680	11/16	3/8	15.500	15 1/2	1.090	1 1/16	11 1/4	1 3/4	1

[a]For application refer to Notes in Table 2.

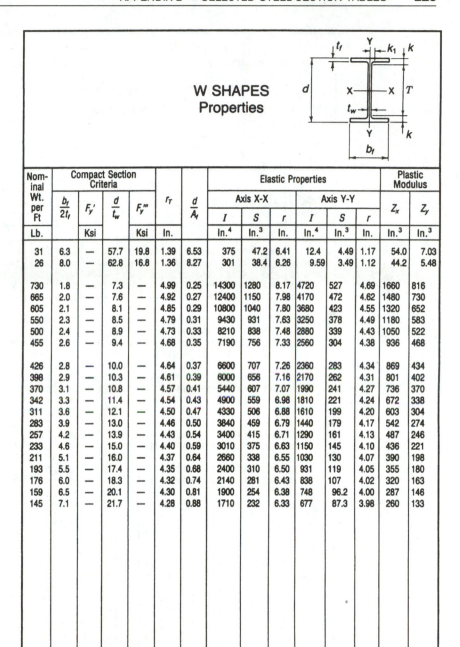

W SHAPES
Properties

Nom-inal Wt. per Ft	Compact Section Criteria				r_T	$\dfrac{d}{A_f}$	Elastic Properties						Plastic Modulus	
	$\dfrac{b_f}{2t_f}$	$F_y{}'$	$\dfrac{d}{t_w}$	$F_y{}'''$			Axis X-X			Axis Y-Y			Z_x	Z_y
							I	S	r	I	S	r		
Lb.		Ksi		Ksi	In.		In.^{4}	In.^{3}	In.	In.^{4}	In.^{3}	In.	In.^{3}	In.^{3}
31	6.3	—	57.7	19.8	1.39	6.53	375	47.2	6.41	12.4	4.49	1.17	54.0	7.03
26	8.0	—	62.8	16.8	1.36	8.27	301	38.4	6.26	9.59	3.49	1.12	44.2	5.48
730	1.8	—	7.3	—	4.99	0.25	14300	1280	8.17	4720	527	4.69	1660	816
665	2.0	—	7.6	—	4.92	0.27	12400	1150	7.98	4170	472	4.62	1480	730
605	2.1	—	8.1	—	4.85	0.29	10800	1040	7.80	3680	423	4.55	1320	652
550	2.3	—	8.5	—	4.79	0.31	9430	931	7.63	3250	378	4.49	1180	583
500	2.4	—	8.9	—	4.73	0.33	8210	838	7.48	2880	339	4.43	1050	522
455	2.6	—	9.4	—	4.68	0.35	7190	756	7.33	2560	304	4.38	936	468
426	2.8	—	10.0	—	4.64	0.37	6600	707	7.26	2360	283	4.34	869	434
398	2.9	—	10.3	—	4.61	0.39	6000	656	7.16	2170	262	4.31	801	402
370	3.1	—	10.8	—	4.57	0.41	5440	607	7.07	1990	241	4.27	736	370
342	3.3	—	11.4	—	4.54	0.43	4900	559	6.98	1810	221	4.24	672	338
311	3.6	—	12.1	—	4.50	0.47	4330	506	6.88	1610	199	4.20	603	304
283	3.9	—	13.0	—	4.46	0.50	3840	459	6.79	1440	179	4.17	542	274
257	4.2	—	13.9	—	4.43	0.54	3400	415	6.71	1290	161	4.13	487	246
233	4.6	—	15.0	—	4.40	0.59	3010	375	6.63	1150	145	4.10	436	221
211	5.1	—	16.0	—	4.37	0.64	2660	338	6.55	1030	130	4.07	390	198
193	5.5	—	17.4	—	4.35	0.68	2400	310	6.50	931	119	4.05	355	180
176	6.0	—	18.3	—	4.32	0.74	2140	281	6.43	838	107	4.02	320	163
159	6.5	—	20.1	—	4.30	0.81	1900	254	6.38	748	96.2	4.00	287	146
145	7.1	—	21.7	—	4.28	0.88	1710	232	6.33	677	87.3	3.98	260	133

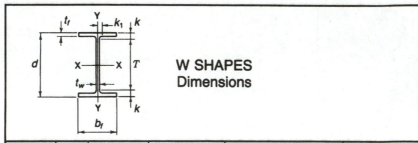

W SHAPES
Dimensions

Desig-nation	Area A	Depth d	Web Thickness t_w	$\frac{t_w}{2}$	Flange Width b_f	Flange Thickness t_f	Distance T	k	k_1
	In.²	In.	In.	In.	In.	In.	In.	In.	In.
W 14×132	38.8	14.66	14⅝	0.645 ⅝	5/16	14.725 14¾	1.030 1	11¼	1 11/16	15/16
×120	35.3	14.48	14½	0.590 9/16	5/16	14.670 14⅝	0.940 15/16	11¼	1⅝	15/16
×109	32.0	14.32	14⅜	0.525 ½	¼	14.605 14⅝	0.860 ⅞	11¼	1 9/16	⅞
× 99	29.1	14.16	14⅛	0.485 ½	¼	14.565 14⅝	0.780 ¾	11¼	1 7/16	⅞
× 90	26.5	14.02	14	0.440 7/16	¼	14.520 14½	0.710 11/16	11¼	1⅜	⅞
W 14× 82	24.1	14.31	14¼	0.510 ½	¼	10.130 10⅛	0.855 ⅞	11	1⅝	1
× 74	21.8	14.17	14⅛	0.450 7/16	¼	10.070 10⅛	0.785 13/16	11	1 9/16	15/16
× 68	20.0	14.04	14	0.415 7/16	¼	10.035 10	0.720 ¾	11	1½	15/16
× 61	17.9	13.89	13⅞	0.375 ⅜	3/16	9.995 10	0.645 ⅝	11	1 7/16	15/16
W 14× 53	15.6	13.92	13⅞	0.370 ⅜	3/16	8.060 8	0.660 11/16	11	1 7/16	15/16
× 48	14.1	13.79	13¾	0.340 5/16	3/16	8.030 8	0.595 ⅝	11	1⅜	⅞
× 43	12.6	13.66	13⅝	0.305 5/16	3/16	7.995 8	0.530 ½	11	1 5/16	⅞
W 14× 38	11.2	14.10	14⅛	0.310 5/16	3/16	6.770 6¾	0.515 ½	12	1 1/16	⅝
× 34	10.0	13.98	14	0.285 5/16	3/16	6.745 6¾	0.455 7/16	12	1	⅝
× 30	8.85	13.84	13⅞	0.270 ¼	⅛	6.730 6¾	0.385 ⅜	12	15/16	⅝
W 14× 26	7.69	13.91	13⅞	0.255 ¼	⅛	5.025 5	0.420 7/16	12	15/16	9/16
× 22	6.49	13.74	13¾	0.230 ¼	⅛	5.000 5	0.335 5/16	12	⅞	9/16

W SHAPES
Properties

Nom-inal Wt. per Ft	Compact Section Criteria				r_T	$\frac{d}{A_f}$	Elastic Properties						Plastic Modulus	
	$\frac{b_f}{2t_f}$	F_y'	$\frac{d}{t_w}$	F_y'''			Axis X-X			Axis Y-Y			Z_x	Z_y
							I	S	r	I	S	r		
Lb.		Ksi		Ksi	In.		In.⁴	In.³	In.	In.⁴	In.³	In.	In.³	In.³
132	7.1	—	22.7	—	4.05	0.97	1530	209	6.28	548	74.5	3.76	234	113
120	7.8	—	24.5	—	4.04	1.05	1380	190	6.24	495	67.5	3.74	212	102
109	8.5	58.6	27.3	—	4.02	1.14	1240	173	6.22	447	61.2	3.73	192	92.7
99	9.3	48.5	29.2	—	4.00	1.25	1110	157	6.17	402	55.2	3.71	173	83.6
90	10.2	40.4	31.9	—	3.99	1.36	999	143	6.14	362	49.9	3.70	157	75.6
82	5.9	—	28.1	—	2.74	1.65	882	123	6.05	148	29.3	2.48	139	44.8
74	6.4	—	31.5	—	2.72	1.79	796	112	6.04	134	26.6	2.48	126	40.6
68	7.0	—	33.8	57.7	2.71	1.94	723	103	6.01	121	24.2	2.46	115	36.9
61	7.7	—	37.0	48.1	2.70	2.15	640	92.2	5.98	107	21.5	2.45	102	32.8
53	6.1	—	37.6	46.7	2.15	2.62	541	77.8	5.89	57.7	14.3	1.92	87.1	22.0
48	6.7	—	40.6	40.2	2.13	2.89	485	70.3	5.85	51.4	12.8	1.91	78.4	19.6
43	7.5	—	44.8	32.9	2.12	3.22	428	62.7	5.82	45.2	11.3	1.89	69.6	17.3
38	6.6	—	45.5	31.9	1.77	4.04	385	54.6	5.87	26.7	7.88	1.55	61.5	12.1
34	7.4	—	49.1	27.4	1.76	4.56	340	48.6	5.83	23.3	6.91	1.53	54.6	10.6
30	8.7	55.3	51.3	25.1	1.74	5.34	291	42.0	5.73	19.6	5.82	1.49	47.3	8.99
26	6.0	—	54.5	22.2	1.28	6.59	245	35.3	5.65	8.91	3.54	1.08	40.2	5.54
22	7.5	—	59.7	18.5	1.25	8.20	199	29.0	5.54	7.00	2.80	1.04	33.2	4.39

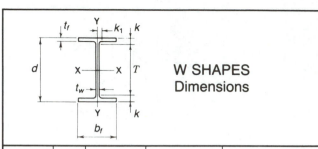

W SHAPES
Dimensions

Desig-nation	Area A	Depth d		Web Thickness t_w		$\dfrac{t_w}{2}$	Flange Width b_f		Flange Thickness t_f		Distance T	k	k_1
	In.²	In.		In.		In.	In.		In.		In.	In.	In.
W 12×336ª	98.8	16.82	16⅞	1.775	1¾	⅞	13.385	13⅜	2.955	2¹⁵⁄₁₆	9½	3¹¹⁄₁₆	1½
×305ª	89.6	16.32	16⅜	1.625	1⅝	¹³⁄₁₆	13.235	13¼	2.705	2¹¹⁄₁₆	9½	3⁷⁄₁₆	1⁷⁄₁₆
×279ª	81.9	15.85	15⅞	1.530	1½	¾	13.140	13⅛	2.470	2½	9½	3³⁄₁₆	1⅜
×252ª	74.1	15.41	15⅜	1.395	1⅜	¹¹⁄₁₆	13.005	13	2.250	2¼	9½	2¹⁵⁄₁₆	1⁵⁄₁₆
×230ª	67.7	15.05	15	1.285	1⁵⁄₁₆	¹¹⁄₁₆	12.895	12⅞	2.070	2¹⁄₁₆	9½	2¾	1¼
×210ª	61.8	14.71	14¾	1.180	1³⁄₁₆	⅝	12.790	12¾	1.900	1⅞	9½	2⅝	1¼
×190	55.8	14.38	14⅜	1.060	1¹⁄₁₆	⁹⁄₁₆	12.670	12⅝	1.735	1¾	9½	2⁷⁄₁₆	1³⁄₁₆
×170	50.0	14.03	14	0.960	¹⁵⁄₁₆	½	12.570	12⅝	1.560	1⁹⁄₁₆	9½	2¼	1⅛
×152	44.7	13.71	13¾	0.870	⅞	⁷⁄₁₆	12.480	12½	1.400	1⅜	9½	2⅛	1¹⁄₁₆
×136	39.9	13.41	13⅜	0.790	¹³⁄₁₆	⁷⁄₁₆	12.400	12⅜	1.250	1¼	9½	1¹⁵⁄₁₆	1
×120	35.3	13.12	13⅛	0.710	¹¹⁄₁₆	⅜	12.320	12⅜	1.105	1⅛	9½	1¹³⁄₁₆	1
×106	31.2	12.89	12⅞	0.610	⅝	⁵⁄₁₆	12.220	12¼	0.990	1	9½	1¹¹⁄₁₆	¹⁵⁄₁₆
× 96	28.2	12.71	12¾	0.550	⁹⁄₁₆	⁵⁄₁₆	12.160	12⅛	0.900	⅞	9½	1⅝	⅞
× 87	25.6	12.53	12½	0.515	½	¼	12.125	12⅛	0.810	¹³⁄₁₆	9½	1½	⅞
× 79	23.2	12.38	12⅜	0.470	½	¼	12.080	12⅛	0.735	¾	9½	1⁷⁄₁₆	⅞
× 72	21.1	12.25	12¼	0.430	⁷⁄₁₆	¼	12.040	12	0.670	¹¹⁄₁₆	9½	1⅜	⅞
× 65	19.1	12.12	12⅛	0.390	⅜	³⁄₁₆	12.000	12	0.605	⅝	9½	1⁵⁄₁₆	¹³⁄₁₆
W 12× 58	17.0	12.19	12¼	0.360	⅜	³⁄₁₆	10.010	10	0.640	⅝	9½	1⅜	¹³⁄₁₆
× 53	15.6	12.06	12	0.345	⅜	³⁄₁₆	9.995	10	0.575	⁹⁄₁₆	9½	1¼	¹³⁄₁₆
W 12× 50	14.7	12.19	12¼	0.370	⅜	³⁄₁₆	8.080	8⅛	0.640	⅝	9½	1⅜	¹³⁄₁₆
× 45	13.2	12.06	12	0.335	⁵⁄₁₆	³⁄₁₆	8.045	8	0.575	⁹⁄₁₆	9½	1¼	¹³⁄₁₆
× 40	11.8	11.94	12	0.295	⁵⁄₁₆	³⁄₁₆	8.005	8	0.515	½	9½	1¼	¾
W 12× 35	10.3	12.50	12½	0.300	⁵⁄₁₆	³⁄₁₆	6.560	6½	0.520	½	10½	1	⁹⁄₁₆
× 30	8.79	12.34	12⅜	0.260	¼	⅛	6.520	6½	0.440	⁷⁄₁₆	10½	¹⁵⁄₁₆	½
× 26	7.65	12.22	12¼	0.230	¼	⅛	6.490	6½	0.380	⅜	10½	⅞	½
W 12× 22	6.48	12.31	12¼	0.260	¼	⅛	4.030	4	0.425	⁷⁄₁₆	10½	⅞	½
× 19	5.57	12.16	12⅛	0.235	¼	⅛	4.005	4	0.350	⅜	10½	¹³⁄₁₆	½
× 16	4.71	11.99	12	0.220	¼	⅛	3.990	4	0.265	¼	10½	¾	½
× 14	4.16	11.91	11⅞	0.200	³⁄₁₆	⅛	3.970	4	0.225	¼	10½	¹¹⁄₁₆	½

ªFor application refer to Notes in Table 2.
Shapes in shaded rows are not available from domestic producers.

W SHAPES
Properties

Nom-inal Wt. per Ft.	Compact Section Criteria				r_T	$\frac{d}{A_f}$	Elastic Properties						Plastic Modulus	
	$\frac{b_f}{2t_f}$	F_y'	$\frac{d}{t_w}$	F_y'''			Axis X-X			Axis Y-Y			Z_x	Z_y
							I	S	r	I	S	r		
Lb.		Ksi		Ksi	In.		In.^{4}	In.^{3}	In.	In.^{4}	In.^{3}	In.	In.^{3}	In.^{3}
336	2.3	—	9.5	—	3.71	0.43	4060	483	6.41	1190	177	3.47	603	274
305	2.4	—	10.0	—	3.67	0.46	3550	435	6.29	1050	159	3.42	537	244
279	2.7	—	10.4	—	3.64	0.49	3110	393	6.16	937	143	3.38	481	220
252	2.9	—	11.0	—	3.59	0.53	2720	353	6.06	828	127	3.34	428	196
230	3.1	—	11.7	—	3.56	0.56	2420	321	5.97	742	115	3.31	386	177
210	3.4	—	12.5	—	3.53	0.61	2140	292	5.89	664	104	3.28	348	159
190	3.7	—	13.6	—	3.50	0.65	1890	263	5.82	589	93.0	3.25	311	143
170	4.0	—	14.6	—	3.47	0.72	1650	235	5.74	517	82.3	3.22	275	126
152	4.5	—	15.8	—	3.44	0.79	1430	209	5.66	454	72.8	3.19	243	111
136	5.0	—	17.0	—	3.41	0.87	1240	186	5.58	398	64.2	3.16	214	98.0
120	5.6	—	18.5	—	3.38	0.96	1070	163	5.51	345	56.0	3.13	186	85.4
106	6.2	—	21.1	—	3.36	1.07	933	145	5.47	301	49.3	3.11	164	75.1
96	6.8	—	23.1	—	3.34	1.16	833	131	5.44	270	44.4	3.09	147	67.5
87	7.5	—	24.3	—	3.32	1.28	740	118	5.38	241	39.7	3.07	132	60.4
79	8.2	62.6	26.3	—	3.31	1.39	662	107	5.34	216	35.8	3.05	119	54.3
72	9.0	52.3	28.5	—	3.29	1.52	597	97.4	5.31	195	32.4	3.04	108	49.2
65	9.9	43.0	31.1	—	3.28	1.67	533	87.9	5.28	174	29.1	3.02	96.8	44.1
58	7.8	—	33.9	57.6	2.72	1.90	475	78.0	5.28	107	21.4	2.51	86.4	32.5
53	8.7	55.9	35.0	54.1	2.71	2.10	425	70.6	5.23	95.8	19.2	2.48	77.9	29.1
50	6.3	—	32.9	60.9	2.17	2.36	394	64.7	5.18	56.3	13.9	1.96	72.4	21.4
45	7.0	—	36.0	51.0	2.15	2.61	350	58.1	5.15	50.0	12.4	1.94	64.7	19.0
40	7.8	—	40.5	40.3	2.14	2.90	310	51.9	5.13	44.1	11.0	1.93	57.5	16.8
35	6.3	—	41.7	38.0	1.74	3.66	285	45.6	5.25	24.5	7.47	1.54	51.2	11.5
30	7.4	—	47.5	29.3	1.73	4.30	238	38.6	5.21	20.3	6.24	1.52	43.1	9.56
26	8.5	57.9	53.1	23.4	1.72	4.95	204	33.4	5.17	17.3	5.34	1.51	37.2	8.17
22	4.7	—	47.3	29.5	1.02	7.19	156	25.4	4.91	4.66	2.31	0.847	29.3	3.66
19	5.7	—	51.7	24.7	1.00	8.67	130	21.3	4.82	3.76	1.88	0.822	24.7	2.98
16	7.5	—	54.5	22.2	0.96	11.3	103	17.1	4.67	2.82	1.41	0.773	20.1	2.26
14	8.8	54.3	59.6	18.6	0.95	13.3	88.6	14.9	4.62	2.36	1.19	0.753	17.4	1.90

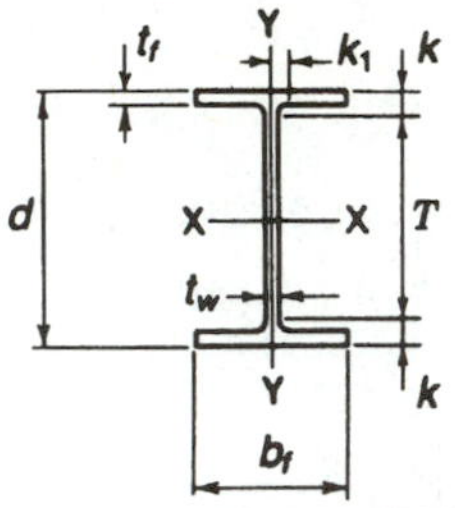

W SHAPES
Dimensions

Designation	Area A	Depth d		Web Thickness t_w		Web $\frac{t_w}{2}$	Flange Width b_f		Flange Thickness t_f		Distance T	Distance k	Distance k_1
	In.^{2}	In.		In.		In.	In.		In.		In.	In.	In.
W 10×112	32.9	11.36	11 3/8	0.755	3/4	3/8	10.415	10 3/8	1.250	1 1/4	7 5/8	1 7/8	15/16
×100	29.4	11.10	11 1/8	0.680	11/16	3/8	10.340	10 3/8	1.120	1 1/8	7 5/8	1 3/4	7/8
× 88	25.9	10.84	10 7/8	0.605	5/8	5/16	10.265	10 1/4	0.990	1	7 5/8	1 5/8	13/16
× 77	22.6	10.60	10 5/8	0.530	1/2	1/4	10.190	10 1/4	0.870	7/8	7 5/8	1 1/2	13/16
× 68	20.0	10.40	10 3/8	0.470	1/2	1/4	10.130	10 1/8	0.770	3/4	7 5/8	1 3/8	3/4
× 60	17.6	10.22	10 1/4	0.420	7/16	1/4	10.080	10 1/8	0.680	11/16	7 5/8	1 5/16	3/4
× 54	15.8	10.09	10 1/8	0.370	3/8	3/16	10.030	10	0.615	5/8	7 5/8	1 1/4	11/16
× 49	14.4	9.98	10	0.340	5/16	3/16	10.000	10	0.560	9/16	7 5/8	1 3/16	11/16
W 10× 45	13.3	10.10	10 1/8	0.350	3/8	3/16	8.020	8	0.620	5/8	7 5/8	1 1/4	11/16
× 39	11.5	9.92	9 7/8	0.315	5/16	3/16	7.985	8	0.530	1/2	7 5/8	1 1/8	11/16
× 33	9.71	9.73	9 3/4	0.290	5/16	3/16	7.960	8	0.435	7/16	7 5/8	1 1/16	11/16
W 10× 30	8.84	10.47	10 1/2	0.300	5/16	3/16	5.810	5 3/4	0.510	1/2	8 5/8	15/16	1/2
× 26	7.61	10.33	10 3/8	0.260	1/4	1/8	5.770	5 3/4	0.440	7/16	8 5/8	7/8	1/2
× 22	6.49	10.17	10 1/8	0.240	1/4	1/8	5.750	5 3/4	0.360	3/8	8 5/8	3/4	1/2
W 10× 19	5.62	10.24	10 1/4	0.250	1/4	1/8	4.020	4	0.395	3/8	8 5/8	13/16	1/2
× 17	4.99	10.11	10 1/8	0.240	1/4	1/8	4.010	4	0.330	5/16	8 5/8	3/4	1/2
× 15	4.41	9.99	10	0.230	1/4	1/8	4.000	4	0.270	1/4	8 5/8	11/16	7/16
× 12	3.54	9.87	9 7/8	0.190	3/16	1/8	3.960	4	0.210	3/16	8 5/8	5/8	7/16

W SHAPES
Properties

Nom-inal Wt. per Ft	Compact Section Criteria				r_T	$\frac{d}{A_f}$	Elastic Properties						Plastic Modulus	
	$\frac{b_f}{2t_f}$	F_y'	$\frac{d}{t_w}$	F_y'''			Axis X-X			Axis Y-Y			Z_x	Z_y
							I	S	r	I	S	r		
Lb.		Ksi		Ksi	In.		In.⁴	In.³	In.	In.⁴	In.³	In.	In.³	In.³
112	4.2	—	15.0	—	2.88	0.87	716	126	4.66	236	45.3	2.68	147	69.2
100	4.6	—	16.3	—	2.85	0.96	623	112	4.60	207	40.0	2.65	130	61.0
88	5.2	—	17.9	—	2.83	1.07	534	98.5	4.54	179	34.8	2.63	113	53.1
77	5.9	—	20.0	—	2.80	1.20	455	85.9	4.49	154	30.1	2.60	97.6	45.9
68	6.6	—	22.1	—	2.79	1.33	394	75.7	4.44	134	26.4	2.59	85.3	40.1
60	7.4	—	24.3	—	2.77	1.49	341	66.7	4.39	116	23.0	2.57	74.6	35.0
54	8.2	63.5	27.3	—	2.75	1.64	303	60.0	4.37	103	20.6	2.56	66.6	31.3
49	8.9	53.0	29.4	—	2.74	1.78	272	54.6	4.35	93.4	18.7	2.54	60.4	28.3
45	6.5	—	28.9	—	2.18	2.03	248	49.1	4.32	53.4	13.3	2.01	54.9	20.3
39	7.5	—	31.5	—	2.16	2.34	209	42.1	4.27	45.0	11.3	1.98	46.8	17.2
33	9.1	50.5	33.6	58.7	2.14	2.81	170	35.0	4.19	36.6	9.20	1.94	38.8	14.0
30	5.7	—	34.9	54.2	1.55	3.53	170	32.4	4.38	16.7	5.75	1.37	36.6	8.84
26	6.6	—	39.7	41.8	1.54	4.07	144	27.9	4.35	14.1	4.89	1.36	31.3	7.50
22	8.0	—	42.4	36.8	1.51	4.91	118	23.2	4.27	11.4	3.97	1.33	26.0	6.10
19	5.1	—	41.0	39.4	1.03	6.45	96.3	18.8	4.14	4.29	2.14	0.874	21.6	3.35
17	6.1	—	42.1	37.2	1.01	7.64	81.9	16.2	4.05	3.56	1.78	0.844	18.7	2.80
15	7.4	—	43.4	35.0	0.99	9.25	68.9	13.8	3.95	2.89	1.45	0.810	16.0	2.30
12	9.4	47.5	51.9	24.5	0.96	11.9	53.8	10.9	3.90	2.18	1.10	0.785	12.6	1.74

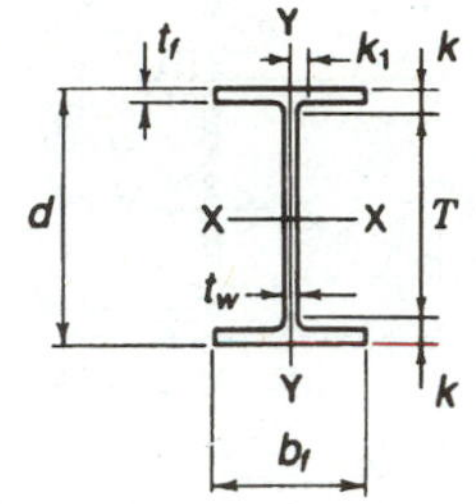

W SHAPES
Dimensions

Designation	Area A	Depth d		Web Thickness t_w		Web $\frac{t_w}{2}$	Flange Width b_f		Flange Thickness t_f		Distance T	Distance k	Distance k_1
	In.^{2}	In.		In.		In.	In.		In.		In.	In.	In.
W 8×67	19.7	9.00	9	0.570	9/16	5/16	8.280	8 1/4	0.935	15/16	6 1/8	1 7/16	11/16
×58	17.1	8.75	8 3/4	0.510	1/2	1/4	8.220	8 1/4	0.810	13/16	6 1/8	1 5/16	11/16
×48	14.1	8.50	8 1/2	0.400	3/8	3/16	8.110	8 1/8	0.685	11/16	6 1/8	1 3/16	5/8
×40	11.7	8.25	8 1/4	0.360	3/8	3/16	8.070	8 1/8	0.560	9/16	6 1/8	1 1/16	5/8
×35	10.3	8.12	8 1/8	0.310	5/16	3/16	8.020	8	0.495	1/2	6 1/8	1	9/16
×31	9.13	8.00	8	0.285	5/16	3/16	7.995	8	0.435	7/16	6 1/8	15/16	9/16
W 8×28	8.25	8.06	8	0.285	5/16	3/16	6.535	6 1/2	0.465	7/16	6 1/8	15/16	9/16
×24	7.08	7.93	7 7/8	0.245	1/4	1/8	6.495	6 1/2	0.400	3/8	6 1/8	7/8	9/16
W 8×21	6.16	8.28	8 1/4	0.250	1/4	1/8	5.270	5 1/4	0.400	3/8	6 5/8	13/16	1/2
×18	5.26	8.14	8 1/8	0.230	1/4	1/8	5.250	5 1/4	0.330	5/16	6 5/8	3/4	7/16
W 8×15	4.44	8.11	8 1/8	0.245	1/4	1/8	4.015	4	0.315	5/16	6 5/8	3/4	1/2
×13	3.84	7.99	8	0.230	1/4	1/8	4.000	4	0.255	1/4	6 5/8	11/16	7/16
×10	2.96	7.89	7 7/8	0.170	3/16	1/8	3.940	4	0.205	3/16	6 5/8	5/8	7/16
W 6×25	7.34	6.38	6 3/8	0.320	5/16	3/16	6.080	6 1/8	0.455	7/16	4 3/4	13/16	7/16
×20	5.87	6.20	6 1/4	0.260	1/4	1/8	6.020	6	0.365	3/8	4 3/4	3/4	7/16
×15	4.43	5.99	6	0.230	1/4	1/8	5.990	6	0.260	1/4	4 3/4	5/8	3/8
W 6×16	4.74	6.28	6 1/4	0.260	1/4	1/8	4.030	4	0.405	3/8	4 3/4	3/4	7/16
×12	3.55	6.03	6	0.230	1/4	1/8	4.000	4	0.280	1/4	4 3/4	5/8	3/8
× 9	2.68	5.90	5 7/8	0.170	3/16	1/8	3.940	4	0.215	3/16	4 3/4	9/16	3/8
W 5×19	5.54	5.15	5 1/8	0.270	1/4	1/8	5.030	5	0.430	7/16	3 1/2	13/16	7/16
×16	4.68	5.01	5	0.240	1/4	1/8	5.000	5	0.360	3/8	3 1/2	3/4	7/16
W 4×13	3.83	4.16	4 1/8	0.280	1/4	1/8	4.060	4	0.345	3/8	2 3/4	11/16	7/16

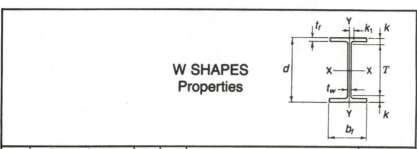

W SHAPES
Properties

Nominal Wt. per Ft	Compact Section Criteria				r_T	$\dfrac{d}{A_f}$	Elastic Properties						Plastic Modulus	
	$\dfrac{b_f}{2t_f}$	F_y'	$\dfrac{d}{t_w}$	F_y'''			Axis X-X			Axis Y-Y			Z_x	Z_y
							I	S	r	I	S	r		
Lb.		Ksi		Ksi	In.		In.^{4}	In.^{3}	In.	In.^{4}	In.^{3}	In.	In.^{3}	In.^{3}
67	4.4	—	15.8	—	2.28	1.16	272	60.4	3.72	88.6	21.4	2.12	70.2	32.7
58	5.1	—	17.2	—	2.26	1.31	228	52.0	3.65	75.1	18.3	2.10	59.8	27.9
48	5.9	—	21.3	—	2.23	1.53	184	43.3	3.61	60.9	15.0	2.08	49.0	22.9
40	7.2	—	22.9	—	2.21	1.83	146	35.5	3.53	49.1	12.2	2.04	39.8	18.5
35	8.1	64.4	26.2	—	2.20	2.05	127	31.2	3.51	42.6	10.6	2.03	34.7	16.1
31	9.2	50.0	28.1	—	2.18	2.30	110	27.5	3.47	37.1	9.27	2.02	30.4	14.1
28	7.0	—	28.3	—	1.77	2.65	98.0	24.3	3.45	21.7	6.63	1.62	27.2	10.1
24	8.1	64.1	32.4	63.0	1.76	3.05	82.8	20.9	3.42	18.3	5.63	1.61	23.2	8.57
21	6.6	—	33.1	60.2	1.41	3.93	75.3	18.2	3.49	9.77	3.71	1.26	20.4	5.69
18	8.0	—	35.4	52.7	1.39	4.70	61.9	15.2	3.43	7.97	3.04	1.23	17.0	4.66
15	6.4	—	33.1	60.3	1.03	6.41	48.0	11.8	3.29	3.41	1.70	0.876	13.6	2.67
13	7.8	—	34.7	54.7	1.01	7.83	39.6	9.91	3.21	2.73	1.37	0.843	11.4	2.15
10	9.6	45.8	46.4	30.7	0.99	9.77	30.8	7.81	3.22	2.09	1.06	0.841	8.87	1.66
25	6.7	—	19.9	—	1.66	2.31	53.4	16.7	2.70	17.1	5.61	1.52	18.9	8.56
20	8.2	62.1	23.8	—	1.64	2.82	41.4	13.4	2.66	13.3	4.41	1.50	14.9	6.72
15	11.5	31.8	26.0	—	1.61	3.85	29.1	9.72	2.56	9.32	3.11	1.46	10.8	4.75
16	5.0	—	24.2	—	1.08	3.85	32.1	10.2	2.60	4.43	2.20	0.966	11.7	3.39
12	7.1	—	26.2	—	1.05	5.38	22.1	7.31	2.49	2.99	1.50	0.918	8.30	2.32
9	9.2	50.3	34.7	54.8	1.03	6.96	16.4	5.56	2.47	2.19	1.11	0.905	6.23	1.72
19	5.8	—	19.1	—	1.38	2.38	26.2	10.2	2.17	9.13	3.63	1.28	11.6	5.53
16	6.9	—	20.9	—	1.37	2.78	21.3	8.51	2.13	7.51	3.00	1.27	9.59	4.57
13	5.9	—	14.9	—	1.10	2.97	11.3	5.46	1.72	3.86	1.90	1.00	6.28	2.92

C

SELECTED AISC COLUMN LOAD TABLES

F_y = 36 ksi
F_y = 50 ksi

COLUMNS
W shapes
Allowable axial loads in kips

Designation		W14							
Wt./ft		193		176		159		145	
F_y		36	50	36	50	36	50	36	50
	0	1227	1704	1119	1554	1009	1401	922	1281
	6	1178	1620	1074	1477	968	1331	884	1217
	7	1167	1603	1064	1461	959	1317	877	1203
	8	1157	1584	1054	1444	950	1301	869	1189
	9	1146	1565	1044	1426	941	1285	860	1174
	10	1134	1545	1034	1407	931	1268	851	1159
	11	1122	1524	1022	1388	921	1250	842	1142
	12	1110	1502	1011	1368	911	1232	832	1125
	13	1097	1479	999	1347	900	1213	822	1108
	14	1083	1455	987	1325	889	1193	812	1090
	15	1069	1431	974	1302	877	1173	801	1071
	16	1055	1406	961	1279	865	1152	790	1051
	17	1040	1380	947	1255	853	1130	779	1031
	18	1025	1353	933	1231	840	1107	767	1011
	19	1010	1326	919	1205	827	1085	755	990
	20	994	1298	904	1179	814	1061	743	968
	22	961	1239	874	1125	786	1012	718	923
	24	927	1178	842	1069	758	960	691	875
	26	891	1113	809	1009	727	906	663	825
	28	853	1046	775	947	696	850	634	773
	30	814	976	739	882	663	791	604	719
	32	774	902	701	815	629	729	573	662
	34	732	826	662	744	594	665	540	603
	36	688	745	622	670	558	598	507	541
	38	643	669	580	601	520	537	472	486
	40	596	604	537	543	480	484	435	438

Effective length in ft KL with respect to least radius of gyration r_y

Properties									
U		2.29	2.29	2.31	2.31	2.32	2.32	2.34	2.34
P_{wo} (kips)		340	473	299	415	251	349	214	298
P_{wi} (kips/in.)		32	45	30	42	27	37	24	34
P_{wb} (kips)		1542	1817	1250	1474	904	1066	688	810
P_{fb} (kips)		467	648	386	536	319	443	267	371
L_c (ft)		16.6	14.1	16.5	14.0	16.4	13.9	16.4	13.9
L_u (ft)		68.1	49.0	62.6	45.0	57.2	41.2	52.6	37.9
A (in.²)		56.8		51.8		46.7		42.7	
I_x (in.⁴)		2400		2140		1900		1710	
I_y (in.⁴)		931		838		748		677	
r_y (in.)		4.05		4.02		4.00		3.98	
Ratio r_x/r_y		1.60		1.60		1.60		1.59	
B_x ⎱ Bending		0.183		0.184		0.184		0.184	
B_y ⎰ factors		0.477		0.484		0.485		0.489	
$a_x/10^6$		358		319		283		255	
$a_y/10^6$		139		125		111		101	
$F'_{ex} (K_x L_x)^2/10^2$ (kips)		438		429		422		416	
$F'_{ey} (K_y L_y)^2/10^2$ (kips)		170		168		166		164	

COLUMNS
W shapes
Allowable axial loads in kips

F_y = 36 ksi
F_y = 50 ksi

Designation		W14								
Wt./ft		132		120		109		99		90	
F_y		36	50	36	50	36	50	36	50†	36	50†
Effective length in ft KL with respect to least radius of gyration r_y	0	838	1164	762	1059	691	960	629	873	572	795
	6	801	1101	729	1002	661	908	600	825	547	751
	7	794	1088	722	990	654	897	595	815	541	742
	8	786	1074	714	977	647	885	589	805	536	732
	9	777	1060	707	963	640	873	582	793	530	722
	10	768	1044	699	949	633	860	575	782	524	711
	11	759	1028	690	935	626	847	568	769	517	700
	12	750	1011	682	919	618	833	561	757	511	689
	13	740	994	673	903	609	818	554	743	504	676
	14	730	976	663	887	601	803	546	730	497	664
	15	719	958	654	870	592	788	538	715	489	651
	16	708	938	644	852	583	772	529	701	482	637
	17	697	919	633	834	574	755	521	685	474	624
	18	686	898	623	815	564	738	512	670	466	609
	19	674	877	612	796	554	721	503	654	458	595
	20	662	856	601	776	544	703	494	637	449	580
	22	637	811	578	735	523	665	475	603	432	548
	24	610	764	554	692	501	626	454	567	413	515
	26	583	714	528	647	478	585	433	529	394	481
	28	554	663	502	599	454	541	411	489	374	444
	30	524	608	475	549	429	496	388	448	353	406
	32	493	551	446	497	403	449	365	404	331	366
	34	461	492	416	443	376	399	340	359	308	325
	36	427	439	385	395	348	356	314	320	285	290
	38	392	394	353	355	319	320	287	288	260	261

Properties										
U	2.47	2.47	2.48	2.48	2.49	2.49	2.50	2.28	2.52	2.29
P_{wo} (kips)	196	272	173	240	148	205	125	174	109	151
P_{wi} (kips/in.)	23	32	21	30	19	26	17	24	16	22
P_{wb} (kips)	587	692	449	529	316	373	249	294	186	220
P_{fb} (kips)	239	332	199	276	166	231	137	190	113	158
L_c (ft)	15.5	13.2	15.5	13.1	15.4	13.1	15.4	13.0	15.3	13.0
L_u (ft)	47.7	34.4	44.1	31.7	40.6	29.2	37.0	26.7	34.0	24.5

A (in.^{2})	38.8	35.3	32.0	29.1	26.5
I_x (in.^{4})	1530	1380	1240	1110	999
I_y (in.^{4})	548	495	447	402	362
r_y (in.)	3.76	3.74	3.73	3.71	3.70
Ratio r_x/r_y	1.67	1.67	1.67	1.66	1.66
B_x } Bending	0.186	0.186	0.185	0.185	0.185
B_y } factors	0.521	0.523	0.523	0.527	0.531
$a_x/10^6$	228.0	204.8	184.5	165.1	148.9
$a_y/10^6$	81.7	73.6	66.3	59.7	54.1
F'_{ex} $(K_x L_x)^2/10^2$ (kips)	409	404	401	395	391
F'_{ey} $(K_y L_y)^2/10^2$ (kips)	147	145	144	143	142

†Flange is noncompact; see discussion preceding column load tables.

F_y = 36 ksi
F_y = 50 ksi

COLUMNS
W shapes
Allowable axial loads in kips

Designation						W14								
Wt./ft	82		74		68		61		53		48		43	
F_y	36	50	36	50	36	50	36	50‡	36	50‡	36	50‡	36‡	50¶
0	521	723	471	654	432	600	387	537	337	468	305	423	272	377
6	482	657	436	595	400	545	358	487	302	408	273	369	244	329
7	474	643	429	581	393	533	351	476	295	395	266	356	237	317
8	465	627	421	567	385	519	345	464	286	380	258	343	230	305
9	456	610	412	552	377	505	338	452	277	364	250	329	223	292
10	446	593	403	536	369	491	330	439	268	348	242	313	215	279
11	435	575	394	520	360	475	322	425	258	330	233	298	207	264
12	425	555	384	502	351	459	314	410	248	312	224	281	199	249
14	402	515	363	465	332	425	297	379	226	273	204	245	181	216
16	377	471	341	426	311	388	278	346	202	229	182	206	161	181
18	351	423	317	383	289	348	258	310	177	184	159	165	140	144
20	323	372	292	337	266	305	237	272	149	149	133	133	117	117
22	293	318	265	287	241	259	214	230	123	123	110	110	96	96
24	261	267	236	241	214	218	190	193	104	104	93	93	81	81
26	227	227	206	206	186	186	165	165	88	88	79	79	69	69
28	196	196	177	177	160	160	142	142	76	76	68	68	60	60
30	171	171	154	154	139	139	124	124	66	66	59	59	52	52
31	160	160	145	145	131	131	116	116	62	62	56	56	49	49
32	150	150	136	136	123	123	109	109	58	58				
34	133	133	120	120	109	109	96	96						
36	119	119	107	107	97	97	86	86						
38	106	106	96	96	87	87	77	77						

Effective length in ft KL with respect to least radius of gyration r_y

Properties

U	3.69	3.69	3.71	3.71	3.75	3.75	3.77	3.43	4.79	4.35	4.39	4.39	4.44	4.44
P_{wo} (kips)	149	207	127	176	112	156	97	135	96	133	84	117	72	100
P_{wi} (kips/in.)	18	26	16	23	15	21	14	19	13	19	12	17	11	15
P_{wb} (kips)	297	350	204	240	160	188	118	139	113	134	88	104	63	75
P_{fb} (kips)	164	228	139	193	117	162	94	130	98	136	80	111	63	88
L_c (ft)	10.7	9.1	10.6	9.0	10.6	9.0	10.6	9.0	8.5	7.2	8.5	7.2	8.4	7.2
L_u (ft)	28.1	20.2	25.9	18.6	23.9	17.2	21.5	15.5	17.7	12.7	16.0	11.5	14.4	10.4

A (in.^{2})	24.1	21.8	20.0	17.9	15.6	14.1	12.6
I_x (in.^{4})	882	796	723	640	541	485	428
I_y (in.^{4})	148	134	121	107	57.7	51.4	45.2
r_y (in.)	2.48	2.48	2.46	2.45	1.92	1.91	1.89
Ratio r_x/r_y	2.44	2.44	2.44	2.44	3.07	3.06	3.08
B_x ⎱ Bending	0.196	0.195	0.194	0.194	0.201	0.201	0.201
B_y ⎰ factors	0.823	0.820	0.826	0.833	1.091	1.102	1.115
$a_x/10^6$	131.4	118.5	107.6	95.4	80.6	71.9	63.6
$a_y/10^6$	22.1	20.0	18.0	16.0	8.6	7.7	6.7
F'_{ex} $(K_x L_x)^2/10^2$ (kips)	380	378	375	371	360	355	351
F'_{ey} $(K_y L_y)^2/10^2$ (kips)	63.8	63.8	62.8	62.2	38.2	37.8	37.0

‡Web may be noncompact for combined axial and bending stress; see AISC ASD Specification Sect. B5.1.
¶Web exceeds AISC ASD Specification Sect. B5.1 noncompact section limit. See discussion preceding tables.
Note: Heavy line indicates Kl/r of 200.

COLUMNS
W shapes
Allowable axial loads in kips

F_y = 36 ksi
F_y = 50 ksi

Designation		W12									
Wt./ft		336		305		279		252		230	
F_y		36	50	36	50	36	50	36	50	36	50
	0	2134	2964	1935	2688	1769	2457	1601	2223	1462	2031
	6	2031	2788	1840	2526	1681	2306	1519	2085	1387	1903
	7	2009	2751	1820	2491	1662	2274	1502	2055	1371	1876
	8	1986	2711	1799	2454	1642	2240	1484	2023	1355	1847
	9	1962	2669	1777	2415	1622	2204	1465	1990	1337	1816
	10	1937	2625	1753	2375	1600	2166	1445	1955	1319	1784
	11	1911	2579	1729	2332	1578	2126	1425	1919	1300	1750
	12	1884	2531	1704	2288	1554	2085	1403	1881	1280	1715
	13	1856	2482	1678	2242	1530	2042	1381	1842	1259	1678
	14	1827	2430	1651	2194	1505	1998	1358	1801	1238	1641
	15	1797	2377	1623	2145	1479	1952	1334	1759	1216	1601
	16	1766	2322	1594	2094	1452	1905	1309	1715	1193	1561
	17	1733	2265	1565	2041	1425	1856	1284	1670	1169	1519
	18	1701	2206	1534	1987	1396	1805	1258	1623	1145	1476
	19	1667	2146	1503	1931	1367	1753	1231	1575	1120	1431
	20	1632	2084	1471	1873	1337	1699	1203	1526	1095	1386
	22	1560	1955	1404	1753	1275	1588	1146	1423	1041	1290
	24	1484	1819	1333	1627	1209	1470	1085	1314	985	1190
	26	1404	1675	1260	1494	1141	1346	1022	1200	927	1084
	28	1321	1525	1183	1354	1069	1216	956	1079	866	972
	30	1235	1366	1102	1206	994	1078	887	952	801	855
	32	1144	1205	1018	1061	916	948	815	837	734	751
	34	1050	1067	930	940	834	839	739	742	664	665
	36	951	952	839	839	749	749	661	661	594	594
	38	854	854	753	753	672	672	594	594	533	533
	40	771	771	679	679	606	606	536	536	481	481

Effective length in ft KL with respect to least radius of gyration r_y

Properties											
U		2.40	2.40	2.41	2.41	2.42	2.42	2.45	2.45	2.46	2.46
P_{wo} (kips)		1178	1636	1005	1396	878	1219	738	1024	636	883
P_{wi} (kips/in.)		64	89	59	81	55	77	50	70	46	64
P_{wb} (kips)		14,480	17,070	11,110	13,100	9274	10,930	7030	8285	5494	6475
P_{fb} (kips)		1965	2729	1646	2287	1373	1907	1139	1582	964	1339
L_c (ft)		14.1	12.0	14.0	11.9	13.9	11.8	13.7	11.6	13.6	11.5
L_u (ft)		107.7	77.5	100.6	72.5	94.5	68.0	87.4	62.9	82.7	59.5

A (in.^{2})	98.8	89.6	81.9	74.1	67.7
I_x (in.^{4})	4060	3550	3110	2720	2420
I_y (in.^{4})	1190	1050	937	828	742
r_y (in.)	3.47	3.42	3.38	3.34	3.31
Ratio r_x/r_y	1.85	1.84	1.82	1.81	1.80
B_x } Bending	0.205	0.206	0.208	0.210	0.211
B_y } factors	0.558	0.564	0.573	0.583	0.589
$a_x/10^6$	605	528	463	405	360
$a_y/10^6$	177	156	139	123	111
F'_{ex} $(K_xL_x)^2/10^2$ (kips)	426	410	393	381	370
F'_{ey} $(K_yL_y)^2/10^2$ (kips)	125	121	118	116	114

F_y = 36 ksi
F_y = 50 ksi

COLUMNS
W shapes
Allowable axial loads in kips

Designation			W12										
Wt./ft		210		190		170		152		136		120	
F_y		36	50	36	50	36	50	36	50	36	50	36	50
Effective length in ft *KL* with respect to least radius of gyration r_y	0	1335	1854	1205	1674	1080	1500	966	1341	862	1197	762	1059
	6	1266	1736	1142	1566	1023	1402	914	1253	815	1117	721	987
	7	1251	1711	1129	1543	1011	1381	903	1233	805	1100	712	972
	8	1236	1684	1115	1518	998	1359	891	1213	795	1082	702	956
	9	1219	1655	1100	1492	984	1335	879	1192	784	1062	692	938
	10	1202	1625	1084	1465	970	1310	866	1169	772	1042	682	920
	11	1185	1594	1068	1437	956	1285	853	1146	760	1021	671	901
	12	1166	1562	1051	1407	940	1258	839	1122	747	999	660	881
	13	1147	1528	1034	1376	924	1230	825	1096	734	976	648	860
	14	1127	1493	1016	1344	908	1200	810	1070	721	952	636	839
	15	1107	1457	997	1311	891	1170	794	1042	707	927	624	817
	16	1086	1419	978	1276	873	1139	778	1014	693	901	611	794
	17	1064	1381	958	1241	855	1107	762	985	678	875	597	770
	18	1042	1341	937	1204	837	1074	745	955	662	848	584	746
	19	1019	1300	916	1167	817	1039	728	924	647	819	569	720
	20	995	1257	894	1128	798	1004	710	892	630	790	555	694
	22	946	1169	849	1047	757	931	673	825	597	730	525	640
	24	894	1076	802	962	714	853	633	754	561	666	493	583
	26	840	977	752	872	668	771	592	680	524	598	460	522
	28	783	874	700	776	621	684	549	601	485	527	425	457
	30	723	766	645	679	571	597	504	524	444	459	388	398
	32	661	673	588	597	519	525	457	461	402	403	349	350
	34	596	596	529	529	465	465	408	408	357	357	310	310
	36	532	532	472	472	415	415	364	364	319	319	277	277
	38	477	477	423	423	372	372	327	327	286	286	248	248
	40	431	431	382	382	336	336	295	295	258	258	224	224

Properties													
U		2.47	2.47	2.49	2.49	2.51	2.51	2.53	2.53	2.55	2.55	2.56	2.56
P_{wo} (kips)		558	774	465	646	389	540	333	462	276	383	232	322
P_{wi} (kips/in.)		42	59	38	53	35	48	31	44	28	40	26	36
P_{wb} (kips)		4255	5014	3084	3635	2291	2700	1705	2010	1277	1505	927	1092
P_{fb} (kips)		812	1128	677	941	548	761	441	613	352	488	275	382
L_c (ft)		13.5	11.5	13.4	11.3	13.3	11.3	13.2	11.2	13.1	11.1	13.0	11.0
L_u (ft)		75.9	54.6	71.2	51.3	64.3	46.3	58.6	42.2	53.2	38.3	48.2	34.7
A (in.^{2})		61.8		55.8		50.0		44.7		39.9		35.3	
I_x (in.^{4})		2140		1890		1650		1430		1240		1070	
I_y (in.^{4})		664		589		517		454		398		345	
r_y (in.)		3.28		3.25		3.22		3.19		3.16		3.13	
Ratio r_x/r_y		1.80		1.79		1.78		1.77		1.77		1.76	
B_x } Bending		0.212		0.212		0.213		0.214		0.215		0.217	
B_y } factors		0.594		0.600		0.608		0.614		0.621		0.630	
$a_x/10^6$		319.5		281.6		245.5		213.4		185.1		159.7	
$a_y/10^6$		99.1		87.8		77.2		67.8		59.4		51.5	
F'_{ex} $(K_xL_x)^2/10^2$ (kips)		360		351		342		332		323		315	
F'_{ey} $(K_yL_y)^2/10^2$ (kips)		112		110		108		106		104		102	

COLUMNS
W shapes
Allowable axial loads in kips

| F_y = 36 ksi |
| F_y = 50 ksi |

Designation		W12										
Wt./ft		106		96		87		79		72		65	
F_y		36	50	36	50	36	50	36	50	36	50	36	50†
	0	674	936	609	846	553	768	501	696	456	633	413	573
	6	637	872	575	788	522	715	473	647	430	589	389	533
	7	629	858	568	775	515	703	467	637	424	579	384	524
	8	620	844	560	762	508	691	460	626	418	569	378	514
	9	611	828	552	748	501	678	453	614	412	558	373	504
	10	602	812	544	733	493	665	446	601	406	547	367	494
	11	593	795	535	718	485	650	439	588	399	535	361	483
	12	583	777	526	701	477	636	431	575	392	522	354	472
	13	572	759	516	685	468	620	423	561	385	509	348	460
	14	561	740	506	667	459	604	415	546	377	496	341	448
	15	550	720	496	649	450	588	407	531	369	482	334	435
	16	539	699	486	630	440	570	398	515	361	468	326	422
	17	527	678	475	611	430	553	389	499	353	453	319	408
	18	514	656	464	591	420	535	379	482	344	438	311	394
	19	502	634	452	570	409	515	370	465	336	422	303	380
	20	489	611	440	549	398	496	360	447	326	406	294	365
	22	462	562	416	505	376	455	339	410	308	372	277	334
	24	433	511	390	458	352	412	317	371	288	336	259	301
	26	404	457	362	408	327	367	294	329	267	297	240	266
	28	372	399	334	356	301	319	270	285	245	258	220	230
	30	340	348	304	310	273	278	245	249	222	225	199	201
	32	305	306	272	273	244	244	219	219	197	197	176	176
	34	271	271	242	242	216	216	194	194	175	175	156	156
	36	241	241	215	215	193	193	173	173	156	156	139	139
	38	217	217	193	193	173	173	155	155	140	140	125	125
	40	196	196	175	175	156	156	140	140	126	126	113	113

Effective length in ft KL with respect to least radius of gyration r_y

Properties												
U	2.59	2.59	2.60	2.60	2.62	2.62	2.63	2.63	2.65	2.65	2.66	2.42
P_{wo} (kips)	185	257	161	223	139	193	122	169	106	148	92	128
P_{wi} (kips/in.)	22	31	20	28	19	26	17	24	15	22	14	20
P_{wb} (kips)	588	693	431	508	354	417	269	317	206	243	154	181
P_{fb} (kips)	221	306	182	253	148	205	122	169	101	140	82	114
L_c (ft)	12.9	10.9	12.8	10.9	12.8	10.9	12.8	10.8	12.7	10.8	12.7	10.7
L_u (ft)	43.3	31.2	39.9	28.7	36.2	26.0	33.3	24.0	30.5	21.9	27.7	20.0

A (in.²)	31.2		28.2		25.6		23.2		21.1		19.1	
I_x (in.⁴)	933		833		740		662		597		533	
I_y (in.⁴)	301		270		241		216		195		174	
r_y (in.)	3.11		3.09		3.07		3.05		3.04		3.02	
Ratio r_x/r_y	1.76		1.76		1.75		1.75		1.75		1.75	
B_x } Bending	0.215		0.215		0.217		0.217		0.217		0.217	
B_y } factors	0.633		0.635		0.645		0.648		0.651		0.656	
$a_x/10^6$	139.1		124.3		110.4		98.6		88.6		79.3	
$a_y/10^6$	45.0		40.1		36.0		32.2		29.1		26.0	
$F'_{ex} (K_x L_x)^2/10^2$ (kips)	310		307		300		296		292		289	
$F'_{ey} (K_y L_y)^2/10^2$ (kips)	100		99.0		97.7		96.5		95.8		94.6	

†Flange is noncompact; see discussion preceding column load tables.

Fy = 36 ksi
Fy = 50 ksi

COLUMNS
W shapes
Allowable axial loads in kips

Designation		W12								
Wt./ft		58		53		50		45		40
F_y		36	50	36	50	36	50	36	50	36	50‡
Effective length in ft KL with respect to least radius of gyration r_y	0	367	510	337	468	318	441	285	396	255	354
	6	341	464	312	425	286	386	256	346	229	309
	7	335	454	307	416	279	374	250	335	223	299
	8	329	443	301	406	271	360	243	322	217	288
	9	322	432	295	395	263	346	235	309	210	276
	10	315	420	288	384	254	331	228	296	203	264
	11	308	407	282	372	246	315	220	281	196	251
	12	301	394	275	360	236	298	211	266	188	237
	13	293	380	268	347	226	281	202	250	180	222
	14	285	365	260	333	216	262	193	233	172	207
	15	276	351	252	319	206	243	183	216	163	191
	16	268	335	244	305	195	223	173	197	154	175
	18	249	302	227	274	171	181	152	159	135	141
	20	230	267	209	241	146	146	129	129	114	114
	22	209	229	189	206	121	121	106	106	94	94
	24	187	193	169	173	102	102	89	89	79	79
	26	164	164	147	147	87	87	76	76	67	67
	28	142	142	127	127	75	75	66	66	58	58
	30	123	123	111	111	65	65	57	57	51	51
	32	108	108	97	97	57	57	50	50	45	45
	34	96	96	86	86						
	38	77	77	69	69						
	41	66	66	59	59						

Properties										
U	3.21	3.21	3.24	2.94	4.10	4.10	4.12	3.75	3.77	3.77
P_{wo} (kips)	89	124	78	108	92	127	75	105	66	92
P_{wi} (kips/in.)	13	18	12	17	13	19	12	17	11	15
P_{wb} (kips)	121	142	106	125	131	155	97	115	66	78
P_{fb} (kips)	92	128	74	103	92	128	74	103	60	83
L_c (ft)	10.6	9.0	10.6	9.0	8.5	7.2	8.5	7.2	8.4	7.2
L_u (ft)	24.4	17.5	22.0	15.9	19.6	14.1	17.7	12.8	16.0	11.5

A (in.^{2})	17.0	15.6	14.7	13.2	11.8
I_x (in.^{4})	475	425	394	350	310
I_y (in.^{4})	107	95.8	56.3	50.0	44.1
r_y (in.)	2.51	2.48	1.96	1.94	1.93
Ratio r_x/r_y	2.10	2.11	2.64	2.65	2.66
B_x } Bending	0.218	0.221	0.227	0.227	0.227
B_y } factors	0.794	0.813	1.058	1.065	1.073
$a_x/10^6$	70.6	63.6	58.8	52.2	46.3
$a_y/10^6$	16.0	14.3	8.4	7.4	6.5
F'_{ex} $(K_xL_x)^2/10^2$ (kips)	289	284	278	275	273
F'_{ey} $(K_yL_y)^2/10^2$ (kips)	65.3	63.8	39.8	39.0	38.6

‡Web may be noncompact for combined axial and bending stress;
  see AISC ASD Specification Sect. B5.1.
Note: Heavy line indicates Kl/r of 200.

COLUMNS
W shapes
Allowable axial loads in kips

F_y = 36 ksi
F_y = 50 ksi

Designation		W10								
Wt./ft		112		100		88		77		68	
F_y		36	50	36	50	36	50	36	50	36	50

Effective length in ft KL with respect to least radius of gyration r_y

KL	112 (36)	112 (50)	100 (36)	100 (50)	88 (36)	88 (50)	77 (36)	77 (50)	68 (36)	68 (50)
0	711	987	635	882	559	777	488	678	432	600
6	663	906	592	808	521	712	454	620	402	548
7	653	888	583	792	513	697	447	607	395	537
8	642	869	573	775	504	682	439	593	388	525
9	631	848	562	756	495	665	431	579	381	512
10	619	827	551	737	485	648	422	564	373	498
11	606	805	540	717	475	630	413	548	365	484
12	593	782	528	696	464	611	404	531	357	469
13	579	757	516	674	453	591	394	513	348	454
14	565	732	503	651	442	571	384	495	339	437
15	550	706	489	627	430	550	373	476	330	421
16	535	679	476	602	417	528	362	457	320	403
17	519	651	461	577	405	505	351	437	310	385
18	503	622	446	550	392	481	339	416	299	366
19	486	591	431	523	378	457	327	394	289	347
20	469	560	416	494	364	432	315	371	278	327
22	433	495	383	435	335	379	289	324	255	285
24	395	425	348	372	304	323	261	275	230	242
26	355	362	312	317	271	275	232	234	204	206
28	313	313	273	273	237	237	202	202	177	177
30	272	272	238	238	206	206	176	176	155	155
32	239	239	209	209	181	181	155	155	136	136
34	212	212	185	185	161	161	137	137	120	120
36	189	189	165	165	143	143	122	122	107	107
38	170	170	148	148	129	129	110	110	96	96
40	153	153	134	134	116	116	99	99	87	87

Properties

	112 (36)	112 (50)	100 (36)	100 (50)	88 (36)	88 (50)	77 (36)	77 (50)	68 (36)	68 (50)
U	2.45	2.45	2.46	2.46	2.49	2.49	2.51	2.51	2.52	2.52
P_{wo} (kips)	255	354	214	298	177	246	143	199	116	162
P_{wi} (kips/in.)	27	38	24	34	22	30	19	27	17	24
P_{wb} (kips)	1388	1636	1014	1196	714	842	480	566	335	395
P_{fb} (kips)	352	488	282	392	221	306	170	237	133	185
L_c (ft)	11.0	9.3	10.9	9.3	10.8	9.2	10.8	9.1	10.7	9.1
L_u (ft)	53.2	38.3	48.2	34.7	43.3	31.2	38.6	27.8	34.8	25.1

	112	100	88	77	68
A (in.^{2})	32.9	29.4	25.9	22.6	20.0
I_x (in.^{4})	716	623	534	455	394
I_y (in.^{4})	236	207	179	154	134
r_y (in.)	2.68	2.65	2.63	2.60	2.59
Ratio r_x/r_y	1.74	1.74	1.73	1.73	1.71
B_x } Bending	0.261	0.263	0.263	0.263	0.264
B_y } factors	0.726	0.735	0.744	0.751	0.758
$a_x/10^6$	106.5	92.7	79.5	67.9	58.7
$a_y/10^6$	35.2	30.8	26.7	22.8	20.0
F'_{ex} $(K_x L_x)^2/10^2$ (kips)	225	219	214	209	204
F'_{ey} $(K_y L_y)^2/10^2$ (kips)	74.5	72.8	71.7	70.1	69.6

F_y = 36 ksi												

F_y = 50 ksi

COLUMNS
W shapes
Allowable axial loads in kips

Designation		W10										
Wt./ft		60		54		49		45		39		33	
F_y		36	50	36	50	36	50	36	50	36	50	36	50
	0	380	528	341	474	311	432	287	399	248	345	210	291
	6	353	482	317	433	289	394	260	351	224	303	189	255
	7	348	472	312	423	284	385	253	340	218	293	184	246
	8	341	461	306	414	279	376	247	328	213	283	179	237
	9	335	450	300	403	273	367	240	316	206	272	173	228
	10	328	437	294	392	268	357	232	303	200	260	167	217
	11	321	425	288	381	262	346	224	289	193	248	161	207
	12	313	412	281	369	256	335	216	274	186	235	155	196
	13	306	398	274	356	249	324	208	259	178	221	149	184
	14	297	383	267	343	242	312	199	243	170	207	142	171
	15	289	368	259	330	235	299	190	227	162	193	135	159
	16	280	353	251	316	228	286	180	209	154	177	127	145
	17	271	337	243	301	221	273	170	191	145	161	120	131
	18	262	320	235	286	213	259	160	172	136	144	112	117
	19	253	303	226	271	205	245	149	154	126	130	103	105
	20	243	285	217	255	197	230	138	139	116	117	95	95
	22	222	248	199	221	180	198	115	115	97	97	78	78
	24	201	209	179	186	161	167	97	97	81	81	66	66
	26	177	178	158	159	142	143	82	82	69	69	56	56
	28	154	154	137	137	123	123	71	71	60	60	48	48
	30	134	134	119	119	107	107	62	62	52	52	42	42
	32	118	118	105	105	94	94	54	54	46	46	37	37
	33	111	111	99	99	88	88	51	51	43	43		
	34	104	104	93	93	83	83						
	36	93	93	83	83	74	74						

Effective length in ft KL with respect to least radius of gyration r_y

Properties												
U		2.55	2.55	2.56	2.56	2.57	2.57	3.25	3.25	3.28	3.28	3.35	3.35
P_{wo} (kips)		99	138	83	116	73	101	79	109	64	89	55	77
P_{wi} (kips/in.)		15	21	13	19	12	17	13	18	11	16	10	15
P_{wb} (kips)		239	282	163	193	127	149	138	163	101	119	79	93
P_{fb} (kips)		104	145	85	118	71	98	86	120	63	88	43	59
L_c (ft)		10.6	9.0	10.6	9.0	10.6	9.0	8.5	7.2	8.4	7.2	8.4	7.1
L_u (ft)		31.1	22.4	28.2	20.3	26.0	18.7	22.8	16.4	19.8	14.2	16.5	11.9

A (in.^{2})	17.6		15.8		14.4		13.3		11.5		9.71
I_x (in.^{4})	341		303		272		248		209		170
I_y (in.^{4})	116		103		93.4		53.4		45.0		36.6
r_y (in.)	2.57		2.56		2.54		2.01		1.98		1.94
Ratio r_x/r_y	1.71		1.71		1.71		2.15		2.16		2.16
B_x } Bending	0.264		0.263		0.264		0.271		0.273		0.277
B_y } factors	0.765		0.767		0.770		1.000		1.018		1.055
$a_x/10^6$	50.5		45.0		40.6		37.2		31.2		25.4
$a_y/10^6$	17.3		15.4		13.8		8.0		6.7		5.4
F'_{ex} $(K_x L_x)^2/10^2$ (kips)	200		198		196		194		189		182
F'_{ey} $(K_y L_y)^2/10^2$ (kips)	68.5		68.0		66.9		41.9		40.7		39.0

Note: Heavy line indicates Kl/r of 200.

INDEX